Cosmetic
Claims Substantiation

COSMETIC SCIENCE AND TECHNOLOGY

Series Editor

ERIC JUNGERMANN

Jungermann Associates, Inc.
Phoenix, Arizona

ADDITIONAL VOLUMES IN PREPARATION

Cosmetic Claims Substantiation

Edited by
Louise B. Aust
Hill Top Research, Inc.
Scottsdale, Arizona

CRC Press
Taylor & Francis Group
Boca Raton London New York

CRC Press is an imprint of the
Taylor & Francis Group, an **informa** business
A TAYLOR & FRANCIS BOOK

CRC Press
Taylor & Francis Group
6000 Broken Sound Parkway NW, Suite 300
Boca Raton, FL 33487-2742

First issued in paperback 2019

ISBN-13: 978-0-8247-9855-0 (hbk)
ISBN-13: 978-0-367-40081-1 (pbk)

Visit the Taylor & Francis Web site at
http://www.taylorandfrancis.com

and the CRC Press Web site at
http://www.crcpress.com

About the Series

The Cosmetic Science and Technology series was conceived to permit discussion of a broad range of current knowledge and theories of cosmetic science and technology. The series is made up of books written either by a single author or edited volumes with a number of contributors. Authorities from industry, academia, and the government are participating in writing these books.

The aim of this series is to cover the many facets of cosmetic science and technology. Topics are drawn from a wide spectrum of disciplines ranging from chemistry, physics, biochemistry, analytical and consumer evaluations to safety, efficacy, toxicity, and regulatory questions. Organic, inorganic, physical and polymer chemistry, emulsion technology, microbiology, dermatology, and toxicology all play a role in cosmetic science.

There is little commonality in the scientific methods, processes, or formulations required for the wide variety of cosmetics and toiletries manufactured. Products range from hair care, oral care, and skin care preparations to lipsticks, nail polishes and extenders, deodorants, body powders and aerosols, to over-the-counter products such as antiperspirants, dandruff treatments, antimicrobial soaps, and acne and sunscreen products.

Cosmetics and toiletries represent a highly diversified field with many subsections of science and "art." Indeed, even in these days of high technology, art and intuition continue to play an important part in the development and evaluation of formulations and the selection of raw materials. There is a move towards more sophisticated scientific methodologies in the fields of preservative efficacy testing, claims substantiation, safety testing, product evaluation, and chemical analyses.

Emphasis in the Cosmetic Science and Technology series is placed on reporting the current status of cosmetic technology and science in addition to historical reviews. The series includes the following titles: *Cosmetic and Drug Preservation; Oral Hygiene Products; Cosmetic Safety; Scientific and Regulatory Foundations of the Cosmetic Industry; A Psychophysical Approach to Cosmetic Product Testing; Selected Analytical Methods and Techniques; Antiperspirants and Deodorants; Glycerine: A Key Cosmetic Raw Material; Clinical Safety and Efficacy Testing; Rheological Properties; Handbook of Cosmetic Microbiology; Sunscreens; Methods of Cutaneous Investigation; Consumer Testing and Evaluation of Personal Care Products* and *Preservative-Free and Self-Preserving Cosmetics and Drugs.* Several of the books have found an international audience and have been translated into Japanese. Contributions range from highly sophisticated and scientific treatises to primers, practical applications, and pragmatic presentations. Authors are encouraged to present their own concepts as well as established theories. Contributors have been asked not to shy away from fields that are still in a state of transition, nor to hesitate to present detailed discussions of their own work. Altogether, we intend to develop in this series a collection of critical surveys and ideas covering diverse phases of the cosmetic industry.

Cosmetic Claims Substantiation, edited by Louise B. Aust, is the eighteenth book published in the Cosmetic Science and Technology series. Claims substantiation, both in the legal and scientific sense, has become an important element in the success or failure of cosmetic products. Obviously, the products per se have to perform and must be chemically stable and well balanced. They must be able to make claims that appeal to the potential consumer, claims that can be substantiated scientifically and can stand the challenge from competitors and governmental regulators. As a result, scientists have developed sophisticated techniques, equipment, and statistical methodologies to support advertising claims that they hope will impress consumers and demonstrate often subtle performance differences to regulatory agencies and networks. We may sometimes wonder whether the application of such complex and complicated technologies to support what often seem to be rather trivial commercials might not be considered as scientific overkill.

In *Cosmetic Claims Substantiation,* experts in the field describe in detail the methods used to demonstrate performance claims for hair and skin care products and describe methods for substantiating such properties as moisturization, mildness, conditioning, cleansing, performance of deodorants and antiperspirants, effectiveness of acne products, and many others. Statistical analyses and the use of consumer testing to confirm product performance, as well as the regulatory aspects of cosmetic claims substantiation, are discussed. This volume covers one of the most important criteria to assure the success of new cosmetic products, namely, the development of claims that the consumer can believe and understand. Of course, the products must be efficacious and safe, but, most important, the message must be sent to the consumer in a believable and provable way.

I want to thank the editor, Louise B. Aust, and all the contributors for collaborating in the editing and writing of this book. Special recognition is also due to Sandra Beberman and the editorial staff at Marcel Dekker, Inc. In addition, I would like to thank my wife Eva, without whose constant support and editorial help I would never have undertaken this project.

Eric Jungermann, Ph.D.
Series Editor

Preface

Advertising, in the form of television, radio, and printed material, is an important facet of marketing consumer products, particularly cosmetics. The cosmetic industry spends hundreds of millions of dollars annually to promote its products. Ads can range from glossy inserts in upscale journals with fragrance strips wafting their latest creation to ads making strong quantitative or comparative claims (e.g., "reduces sweat by 50%," or "more antimicrobial action than other brands," and so forth).

The first set of ads falls into the category of "buy me please, because you like me or because I smell good;" they require no substantiation, only good aesthetics. The second set requires proof and scientific substantiation. Agencies such as the FTC and FDA, National Advertising Division (NAD) and, above all, the competitors monitor such claims and make sure that proper substantiation exists. Claims and claims substantiation have become a major activity in the cosmetic industry and involve scientists, lawyers, regulatory agencies, and marketing managers to push for more and more outrageous claims.

In the field of cosmetics and personal care products, claims substantiation is a major priority. New testing methodologies are often involved and require the development of highly sophisticated and innovative technologies and instrumental techniques. Research and development groups work diligently to create and promote the best possible products.

Since consumer awareness of our products is largely determined by labeling and advertising, it is in the cosmetic industry's best interest to develop and substantiate the strongest and most compelling claims possible. However, as all technical personnel should be aware, the basic performance of a product is dictated by its formula, mode of application, and instructions for use. Even the best designed study protocol cannot make a product efficacious if the synergistic properties of the product are not forthcoming.

The efficacy of commercial products is generally measured against standard conditions such as an untreated control or internal standard. Often the claims for a cosmetic product may be in a new or preemptive context. When warranted, the relative efficacy of the product may be determined on a competitive basis. In many of these cases, the claims substantiation data may also reflect new testing methodologies. In conjunction with this technical evaluation, claims substantiation is also developed with consumer input through market testing and user evaluations.

The following chapters are a compilation of efforts from scientists who work with new products and claims to support these products on a daily basis. Methods for evaluating the efficacy of many different product types, including moisturizers, deodorants, cleansing agents, and cosmetics are described in detail. Furthermore, the handling of statistical analysis of claims data and the legal ramifications that loom in the distance are also discussed. Experimental design, procedures and statistical analyses are among the critical components of a study whose objective is to support a claim. How a study is designed, implemented, and analyzed can affect the final results. Proper study controls and product application procedures also play an important role in the study's outcome. Experts in this arena for various product areas have contributed to this book. It should serve as a resource for both the novice and the seasoned clinician in designing studies.

In the following chapters, a thorough analysis of cosmetic claims substantiation will be presented for the cosmetic chemist, investigative dermatologist, and all individuals working in the cosmetics industry. We will explore types of testing for claims substantiation as well as how to evaluate specific product types. Regulatory aspects of these methods will also be discussed.

Developing methods for claims support is the important work of many of our authors, and I am deeply grateful for their efforts. I am particularly indebted to my mentor, friend, and series editor, Eric Junger-mann, for his consistent support and guidance. We all agree that the methods are only as good as the products, and that the consumers who ultimately use these products will be the collective judge.

Louise B. Aust

Contents

Contributors

Louise B. Aust Hill Top Research, Inc., Scottsdale, Arizona

James D. Ayres Artistry Product Development, R&D, Amway Corporation, Ada, Michigan

James P. Bowman Hill Top Research, Inc., Miamiville, Ohio

Gail Vance Civille Sensory Spectrum, Chatham, New Jersey

Jennifer B. Davis Hyman, Phelps & McNamara, P.C., Washington, D.C.

Clare A. Dus Sensory Spectrum, Chatham, New Jersey

Al Gabbianelli Helene Curtis, Rolling Meadows, Illinois

Maximo C. Gacula, Jr. Gacula Associates, Scottsdale, Arizona

Howard I. Maibach Department of Dermatology, University of California, San Francisco, California

Kenneth D. Marenus Research and Development, Estee Lauder Research Laboratories, Melville, New York

Rolf Mast International Beauty Design, Riverside, California

Stephen H. McNamara Hyman, Phelps & McNamara, P.C., Washington, D.C.

Norman Meltzer Helene Curtis, Rolling Meadows, Illinois

Otto H. Mills, Jr. Hill Top Research, Inc., East Brunswick, New Jersey

Richard I. Murahata Clinical and Appraisal Science, Unilever Research U.S., Edgewater, New Jersey

Gregg A. Nicoll Consumer Science, Unilever Research U.S., Edgewater, New Jersey

Linda P. Oddo Hill Top Research, Inc., Scottsdale, Arizona

Amy Qualls   Helene Curtis, Rolling Meadows, Illinois

Steven Rachui   Stephens and Associates, Inc., Carrolton, Texas

Judith Rahn   Helene Curtis, Rolling Meadows, Illinois

Marie Rudolph   Sensory Spectrum, Chatham, New Jersey

Jagbir Singh   Temple University, Philadelphia, Pennsylvania

Elaine Stern   Helene Curtis, Rolling Meadows, Illinois

John E. Wild   Hill Top Research, Inc., Miamiville, Ohio

Mark Willoughby   University of California, San Francisco, California

Grace Yang   Helene Curtis, Rolling Meadows, Illinois

1

Regulatory Aspects of Cosmetic Claims Substantiation

Jennifer B. Davis and Stephen H. McNamara
Hyman, Phelps & McNamara, P.C., Washington, D.C.

INTRODUCTION

When marketing a cosmetic product, it is important to consider whether the statements made about the product on its label, on other packaging and labeling, and in advertising comport with applicable laws and regulations. This chapter reviews the regulatory requirements for substantiation of cosmetic claims, providing pertinent examples and highlighting recent enforcement trends.

FEDERAL AUTHORITY OVER CLAIMS IN LABELING AND ADVERTISING OF COSMETIC PRODUCTS

The primary sources of federal authority over cosmetic claims are the Federal Food, Drug, and Cosmetic Act (FDC Act) (1), the Fair Packaging and Labeling Act (FPLA) (2), and the Federal Trade Commission Act (FTC Act) (3). These are national laws, enacted by Congress and signed by the president of the United States. The FDC Act and the FPLA are primarily concerned with labeling, while the FTC Act's provisions focus on advertising.

Responsibility for enforcing these statutes is delegated to the Food and Drug Administration (FDA), an agency of the Department of Health and Human Services (DHHS), and the Federal Trade Commission (FTC). The FDA enforces the FDC Act and the FPLA, and has primary jurisdiction over labeling (4). The FTC enforces the FTC Act and is primarily concerned with advertising. The division of authority between the FDA and FTC has been further clarified by a 1954 Memorandum of Understanding (MOU) between the two agencies, which remains in effect today (5). This liaison agreement establishes that the FTC will assume primary responsibility respecting the "truth or falsity of all advertising (other than labeling) of . . . cosmetics," while FDA will retain jurisdiction over "all matters regulating the labeling of . . . cosmetics." The agencies also agree to consult each other and to collaborate when similar claims are found in both the labeling and advertising of a product or when claims could be construed as either advertising or labeling based on the circumstances of product distribution (6).

FDA REGULATION

Definition of *Cosmetic*

The FDC Act defines the term *cosmetic* as:

> (1) articles intended to be rubbed, poured, sprinkled, or sprayed on, introduced into, or otherwise applied to the human body or any part thereof for cleansing, beautifying, promoting attractiveness, or altering the appearance, and (2) articles intended for use as a component of any such articles; except that such term shall not include soap (7).

The same definition is incorporated by reference in the FPLA (8). All articles that fall within this definition are subject to regulation as "cosmetics."

Prohibition of False or Misleading Labeling

In addition to regulations respecting ingredients (9), color additives (10), safety substantiation (11), and the information required to be listed on product labels (12), all cosmetics are subject to the general prohibition against false or misleading labeling. Pursuant to section 602(a) of the FDC Act, a cosmetic is deemed to be misbranded "if its labeling is false or misleading in any particular" (13). In determining whether a product label is misleading, section 201(n) of the FDC Act directs the FDA to consider

> not only representations made or suggested by statement, word, design, device, or any combination thereof, but also the extent to which the labeling . . . fails to reveal facts material in the light of such representations or material with respect to consequences which may result from the use of the article. . . . (14)

One way in which a cosmetic product can be misbranded and consequently subject to FDA enforcement is if the label contains a false, misleading, or unsubstantiated claim about the benefits of the product that is deceptive to the reasonable consumer. For example, a cosmetic that claims on its label to contain a particular desirable ingredient but that does not in fact contain the substance would be misbranded pursuant to section 602(a). A reasonable consumer relying on the claim in selecting and purchasing the product would suffer an economic loss as a result of the manufacturer's false and misleading promotional label statement.

FDA Enforcement Authority and Activities

If the FDA believes that a cosmetic product is misbranded because of false or misleading labeling, it has several informal and formal enforcement remedies at its disposal to bring about correction of the problem and to protect consumers from violative products. Usually, the first measure the FDA employs is the issuance of a warning letter to the responsible company, encouraging voluntary correction of the problem. A warning letter notifies the responsible individual and/or company that the agency believes one or more of its products or activities is in violation of the FDC Act or other acts enforced by the FDA and states that failure to correct the violation promptly may result in immediate administrative and/or regulatory enforcement action without further notice (15). However, the FDA is not required to issue a warning letter prior to pursuing other enforcement remedies.

Generally, the recipient of a warning letter is given 15 days to respond in writing with a description of the steps that are being taken to correct the violation and an estimate of when the correction will be completed. The recipient also may be asked to provide follow-up documentation that a promised correction has been made (16).

In addition, the FDA may request a manufacturer to recall a cosmetic product voluntarily if "a product that has been distributed presents a risk of illness or injury or gross consumer deception" (17). Recall requests are reserved for "urgent" situations and are not used by the FDA for minor violations that do not present concerns with respect to safety or fraud (18). If a company agrees to recall its product, the agency works with the company to develop and implement an appropriate recall strategy (19).

When the FDA believes that a serious violation has occurred that is not likely to be addressed adequately or expediently by a recall or other voluntary corrective action on the part of the responsible company, the agency can ask a U.S. attorney to initiate a civil seizure action against the product (20), an injunction to prevent the company from distributing the product (21), or a criminal prosecution of the company and/or responsible individuals in a U.S. district court (22). Violations of the FPLA, however, are not subject to criminal prosecution (23).

Because cosmetics generally present fewer risks than other products that the FDA regulates, the agency devotes a relatively small portion of its resources to cosmetic regulation. Most of that allotment is directed toward safety-related concerns instead of economic concerns presented by labeling claims. As the Advisory Committee on the Food and Drug Administration stated in its 1991 Final Report, "It is well known . . . that the Agency has for several years largely abandoned efforts to combat economic deception in the sale of . . . cosmetics, choosing appropriately to allocate depleted resources to safety-related violations" (24). More recently, the acting director of the FDA's Office of Cosmetics and Colors explained in a presentation to the Society of Cosmetic Chemists that "[w]hile FDA is concerned that cosmetic labeling is not false and misleading, product claims are not subject to premarket review and approval, [and] we take into account a certain amount of promotional zeal ('puffery')" (25).

The statistics are even more telling. For example, in 1992, the FDA conducted 578 inspections of cosmetic manufacturing facilities compared to 6211 inspections of food product manufacturers (26). In 1993, the reported number of cosmetic facility inspections dropped to 227 compared to 5900 inspections of food manufacturers (27). The agency estimates that even fewer inspections of cosmetic manufacturing facilities were conducted in 1994 and 1995, respectively (28). Consistent with this trend of low-level activity concerning cosmetic products, in 1992 the FDA initiated only two seizure actions against cosmetic products compared with 81 for foods (29), and in 1993 the agency initiated only 6 recalls against cosmetic products compared to 655 recalls of food items (30).

Cosmetic Drugs

At present, the only truly significant area of FDA enforcement activity involving cosmetic claims that do not present a safety-related concern is the situation in which the promotional statements trigger categorization and regulation of the cosmetic product as a drug. Section 201(g) of the FDC Act defines a "drug" as "articles intended for use in the . . . cure, mitigation, treatment, or prevention of disease in man . . . and . . . articles (other than food) intended to affect the structure or any function of the body of man . . ." (31).

The "intended use" or intended effect of an article may be determined by reference to the claims made in labeling and advertising for it. Cosmetic effects are considered to be superficial, while drug effects are thought to be more physiological in nature. Two cases from the 1960s involving skin care products illustrate the types of cosmetic product claims that have been found to trigger drug status.

United States v. An Article . . . Labeled in Part: Sudden Change (32) was a civil seizure action against an albumin protein cosmetic product named "Sudden Change." The product acted to smooth and firm the skin temporarily by tightening the surface as it dried. Displayed on the label of Sudden Change were claims that the product "gives a face lift without surgery," "lifts puffs under the eyes," and "provides a tingling sensation to indicate the article is working" (33). Although the product did not contain any pharmacologically active ingredients, nevertheless, the reviewing court held that so long as these kinds of claims appeared on the label, Sudden Change would be deemed a drug within the meaning of the FDC Act. In *United States v. An Article . . . Labeled in Part: Line Away Temporary Wrinkle Smoother, Coty* (34), a similar product named "Line Away" was deemed to be subject to regulation as a drug based on labeling statements that claimed the product "is not a face lift, not a treatment," "contains absolutely no harmful chemicals, no hormones," but produces "a tingling sensation . . . [which] means that Line Away is at work—smoothing, firming, tightening" (35).

Working with the skin cream example, if a product is claimed to have the physiological effect of eliminating wrinkles, it is subject to regulation as a drug because it has the "intended use" of affecting the structure of the skin. On the other hand, if a skin cream claims only to soften, smooth, or visibly reduce the "appearance" or "look" of wrinkles, it is not likely to trigger drug status.

During the late 1980s, the FDA issued a number of regulatory letters to various companies that were selling "antiaging" and "antiwrinkle" products. Among the types of claims made for those products were statements that the articles enhanced revitalizing cellular activity, repaired and restored the skin, and stimulated the production of essential skin proteins. The FDA asserted that these claims created drug status for the products and that the products were "new drugs" that were not the subject of approved new drug applications—and were therefore illegal.

"Drug" status has significant consequences for a would-be cosmetic. Drug manufacturers must comply with requirements for establishment registration and product listing with the FDA (36), manufacture products in accordance with "current good manufacturing practice" regulations (37), and even obtain prior FDA approval before marketing if a drug product is a "new drug" pursuant to section 201(p) of the FDC Act (38).

FTC REGULATION

Definition of *Cosmetic*

Section 15 of the FTC Act utilizes the same definition of *cosmetic* as the FDC Act, reviewed above (39).

Prohibition of Unfair or Deceptive Acts or Practices in or Affecting Commerce

Section 5 of the FTC Act prohibits "unfair or deceptive acts or practices in or affecting commerce" (40). A demonstrably false advertisement constitutes an "unfair or deceptive act or practice" (41).

Section 12 of the FTC Act contains additional prohibitions specifically applicable to food, drugs, devices, and cosmetics (42). It is "unlawful for any person, partnership or corporation to disseminate, or cause to be disseminated, any false advertisement . . . for the purpose of inducing, or which is likely to induce, directly or indirectly the purchase of food, drugs, devices, services, or cosmetics . . ." (43). "The dissemination or the causing to be disseminated of any false advertisement" constitutes an "unfair or deceptive act or practice in or affecting commerce within the meaning of section [5]" of the FTC Act (44). Curiously, these specific prohibitions have had little significance in FTC cases to date (45).

Section 15(a) (1) of the FTC Act defines a "false advertisement" as

> an advertisement, other than labeling, which is misleading in a material respect; and in determining whether any advertisement is misleading, there shall be taken into account (among other things) not only representations made or suggested by statement, word, design, device, sound, or any combination thereof, but also the extent to which the advertisement fails to reveal facts material in the light of such representations or material with respect to consequences which may result from the use of the commodity to which the advertisement relates. . . . (46)

This language is almost identical to the FDC Act's provision that prescribes how to determine whether product labeling is misleading for purposes of the prohibition on misbranding (47). An advertisement may be deemed false or deceptive for what it says or what it fails to say about a product. An advertisement also may be deemed false or deceptive for what it says about a product as compared to similar competing products.

FTC Enforcement Authority and Activities

Like the FDA, the FTC has several informal and formal enforcement remedies to curtail false or misleading advertisements and advertising campaigns. If the FTC believes an advertisement is deceptive or unsubstantiated in violation of the FTC Act, it may initiate an informal investigation. Usually, FTC staff will request voluntary submission of all advertising materials bearing on substantiation for the claims in question. Through voluntary cooperation on the part of the company, misunderstandings and minor problems can usually be corrected without more formal government action.

If a company refuses to respond to a request for the "voluntary" submission of information, the FTC can issue a formal Civil Investigative Demand for the materials, which is enforceable in a federal court (48).

When the FTC believes that a serious violation of the law has occurred and that it is in the best interest of consumers for the commission to intervene, the commission can issue a formal complaint against the company (49). "Where time, the nature of the proceeding, and the public interest permit," however, the FTC must allow the offending company to submit a proposed consent agreement (50). If the FTC chooses to initially accept a consent agreement, it is published for comment in the *Federal Register* (51). After 60 days, the FTC may then officially accept the agreement as a binding consent order (52).

Consent orders usually contain prohibitions on company practices questioned during an investigation, prohibitions on possible future violations, and a description of any remedies to which the parties have agreed. Remedial measures may include corrective advertising, restitution, or affirmative disclosures. After the order is finalized, a company is usually given 60 days to file a compliance report detailing the steps it will take to correct the violations (53).

If an agreement cannot be reached, the FTC will issue its complaint and the case will be litigated before an administrative law judge (54). Popular remedies available to the FTC in this case include the cease-and-desist order, affirmative disclosure, and corrective advertising. A cease-and-desist order may altogether prohibit a manufacturer from making a particular claim or require that the advertiser substantiate the claim. An order of affirmative disclosure directs a company to add to future advertising omitted material information, the absence of which causes a claim to be deceptive. Corrective advertising requires a manufacturer to state affirmatively that the product does not actually produce a previously claimed effect (55). Pursuant to section 19 of the FTC Act; the FTC can also seek consumer relief (56). In this case, a manufacturer may be ordered to refund money to consumers on the theory that they might have been induced to purchase the product based on its deceptive claim. Final FTC orders are reviewable and enforceable in the federal courts (57).

One of the FTC's most effective tools for preventing deceptive advertising is the requirement that an advertiser have a "reasonable basis" to substantiate its advertising claims. An advertiser's failure to possess and rely on a reasonable basis for its advertising claims constitutes a violation of the FTC Act (58). The type, quantity, and quality of substantiation required depends on the particular claim, product, and advertiser (59).

In evaluating the propriety of advertising claims, the FTC considers the overall impression created by an ad as well as how the reasonable consumer would perceive the claim in the context of the advertisement (60). Once the FTC identifies a particular claim that an ad intends to convey, the commission then determines the "reasonable basis" necessary to substantiate the claim. In so doing, the FTC distinguishes between various types of claims.

Statements that reference the support that an advertiser has for a claim are called "establishment claims." An establishment claim can be explicit, using a phrase such as "doctors recommend," "tests prove," or "studies show"; or it can be implicit, employing a visual aid or language that suggests that the claim is based on scientific evidence (e.g., someone wearing a white lab coat and glasses). For establishment claims that are "specific," meaning that they reference a type or quantity of substantiation, the advertiser must be in possession of that type or quantity of substantiation to support the claims. For example, if an advertiser states that "three salon studies prove," there must be three salon studies and they must be of sufficient quality to constitute a "reasonable basis" for the claim.

For nonestablishment claims (and nonspecific establishment claims), the FTC determines the type of substantiation required by balancing several factors. These factors include the kind of claim and product, the perception of the reasonable consumer, the consequences of a false claim and the benefits of a truthful claim, the cost of developing substantiation for the claim, and the amount of substantiation that experts believe is reasonable (61).

The general rule is that an advertiser must, at the time a claim is made, have in its possession reliable and competent substantiating data of the type and quantity appropriate for the representation. "Reliable and competent" data mean "tests, analyses, research, studies, or other evidence conducted and evaluated in an objective manner by persons qualified to do so, using procedures generally accepted in the profession or science to yield accurate and reliable results" (62). In addition, if the substantiating data are subject to some limitation, this qualification must be made apparent to the consumer (63). For example, if an advertisement claims that a cream is proven to fade skin discoloration but this result is not achieved until after the product has been used for 3 months, this would be an important qualification that should be disclosed to consumers.

Manufacturers of certain cosmetic-type over-the-counter (OTC, i.e., nonprescription) drug products, such as sunscreens and antiperspirants, may make claims that are approved in the FDA's OTC drug monographs. The FTC accepts the conclusions and recommendations of the FDA based upon the agency's OTC Advisory Review Panels as evidence of adequate substantiation for OTC drug performance claims, and the Commission has allowed manufacturers to rely on FDA final monographs for such substantiation (64). Tentative final monographs are also presumed reliable for claims substantiation (65).

Traditionally, cosmetic advertising has relied heavily on "puffery" and subjective claims. The FTC generally does not consider "mere puffing" to be a violation of the FTC Act. Words such as *easy, perfect, amazing, wonderful*, and *excellent* are regarded as mere puffing (66). In addition, the FTC does not require substantiation of "subjective" claims about taste, smell, appearance, or feel. For example, the FTC would regard claims that a product "beautifies your skin," "freshens like an ocean breeze," or "smooths and silkens the skin" as nonactionable.

There have not been many recent FTC actions involving advertising claims for cosmetic products. Like the FDA, the FTC has a tendency not to pursue enforcement actions against cosmetic advertising unless the use of a product in accordance with its advertising claims presents a risk to public health or safety or unless the claim approaches being fraudulent. Most of the notable FTC cases involving deceptive cosmetic claims occurred in the 1940s, 1950s, and 1960s and involved products claimed to cure baldness (67), to color hair permanently (68), or to remove hair permanently (69). In the 1980s, the FTC began to act against "antiaging" claims, but it abandoned this pursuit after the FDA instituted industrywide regulatory action in 1987 (70).

In 1989, the FTC filed a complaint against Revlon regarding its advertising for "Ultima II ProCollagen Anti-Cellulite Body Complex." Among the claims Revlon made were that "no woman has to resign herself to unattractive ripples, bumpy texture, and slackness caused by cellulite" and that its product "increases skin circulation to help disperse toxins and excess water that contribute to cellulite pockets, and ... builds sub-skin tissue strength and tone for smoother support" (71).

In 1994, the FTC brought action against another purported cellulite reduction treatment named Silueta Sistema, which was advertised primarily on Spanish-language television. One of the ads claimed that "Silueta is an astonishing treatment in two steps which penetrates the skin and attacks and dissolves the fat cells which are the cause of those ugly cellulite bumps, and later expels them from your body." The system comprised a moisturizer and diuretic tablets, which 63% of consumers returned due to dissatisfaction. The reviewing court ordered the manufacturer (1) to cease and desist from making all such claims for the product unless supported by competent and reliable scientific evidence, (2) to refund fully all of its customers' money, and (3) to submit reports to the FTC for a period of 5 years detailing the company's efforts to comply with the court order as well as any change in the corporate structure of the company (72).

THE NATIONAL ADVERTISING DIVISION OF THE COUNCIL OF BETTER BUSINESS BUREAUS

A significant source of nongovernmental advertising scrutiny that can affect a company's ability to make certain claims for its cosmetic products is the National Advertising Division of the Council of Better Business Bureaus (NAD). This is a self-regulatory body established in 1971 by the American Advertising Federation, the American Association of Advertising Agencies, the Association of National Advertisers, and the Council of Better Business Bureaus to evaluate challenges to the truth and accuracy of advertising (73). With the recent lack of FDA and FTC enforcement for cosmetic products, the NAD has become the major body to regulate the cosmetic world.

The NAD functions much like a judicial body conducting investigations of suspect advertising, defining issues, collecting and evaluating data, and determining whether claims are adequately substantiated. While some investigations arise from NAD's own scrutiny of television, radio and print advertising, most cases are initiated by consumers, competitors, and professional associations (74).

When NAD receives a complaint that appears to present a reasonable question about the validity of a claim in an advertisement, it sends a letter to the advertiser citing the claim(s) that must be substantiated and requesting examples of similar claims currently appearing in national advertising. If, after reviewing the information, NAD decides that the challenged claims are adequately supported, it notifies the challenger and terminates the investigation. On the other hand, if NAD determines that substantiation is insufficient to support the questioned claims, it requests that the offending advertiser correct the claims or refrain from using such claims in future advertising (75).

If a challenged advertiser disagrees with NAD's findings or if the controversy cannot be resolved, the case may be appealed to the National Advertising Review Board (NARB). The NARB comprises 50 representatives from advertising companies, advertising agencies, and the public sector. The NARB chairman selects an impartial five-member panel to hear and decide each appeal (76).

If an advertiser refuses altogether to participate in NAD's self-regulatory process, demonstrates inadequate cooperative efforts, or continually engages in deceptive advertising practices, NAD may decide that its self-regulatory process is not a proper forum for problem solving. In that case, with the concurrence of the NARB chairman, NAD may file a complaint with the appropriate government agency, such as the FTC (77).

The NAD publishes its decisions monthly in *NAD Case Reports*. Each report describes the advertiser, product, and basis for investigation as well as the advertiser's response and the case resolution. The reports pertaining to cosmetic products can be extremely helpful to a cosmetic manufacturer in predicting the type and degree of substantiation necessary to support a proposed claim.

Generally speaking, NAD is more concerned with the quality of studies cited in support of a claim than with the quantity of such data. In addition, NAD is more likely than the FTC to raise concern about claims that might be regarded as "subjective" or "puffery." The following recent NAD cases illustrate the typical focus of a NAD inquiry with respect to cosmetic claims.

Through its regular advertisement monitoring activities, NAD identified as potentially deceptive a Christian Dior Perfumes, Inc., print ad for Dior Svelte Cellulite Control Complex depicting the lower half of a woman with a red sash tied around her hips. The text of the advertisement included the following claims: (1) "A unique complex containing . . . plant extracts that . . . regulate lipid balance and 'streamline' the appearance of body contours," (2) "fast acting ultra-penetrating gel that works effectively *without massage*," (3) "You'll see visible improvement in a matter of weeks," and (4) "A firmer silhouette" (78). To support the first claim, Christian Dior submitted the results of cell-culture tests performed on the primary ingredients. To support the efficacy claims, the company cited a blind study comparing its product to a competing product that measured cellulite reduction in 128 women over a 2-month period. In support of the fourth claim, the company described the subjective assessment of an expert who evaluated the suppleness, firmness, and appearance of the women's skin after 1 and 2 months. The NAD determined that this documentation provided a reasonable basis for Christian Dior's claims (79).

In another case, a consumer wrote to NAD challenging the Elizabeth Arden Company's advertisement for Alpha-Ceramide Intensive Skin Treatment. The ad depicted the four-step product, claimed that it "could improve the overall quality of your skin (including a reduction in the appearance of fine lines, wrinkles and uneven skin tone) on an average of 42% and up to 68% as judged by a dermatologist," and described the product as a "unique progressive alpha-hydroxy system" (80). The challenger questioned whether the company had any objective effectiveness data besides the paid, subjective opinion of a dermatologist. Elizabeth Arden Company submitted two double-blind studies conducted by independent, respected laboratories. Both tests demonstrated statistical significance of at least the 95% confidence level. In addition, the company cited a 3-month consumer perception study of the product's efficacy that was consistent with the observations of the previous studies. The NAD concluded that "the clinical data submitted by the advertiser [are] independent, objective, well-controlled, and well within what appear to be, in our experience, industry-accepted standards." Thus, NAD determined that the print ad claims were adequately substantiated (81).

A third case illustrates NAD's position on comparative claims alleging product superiority. Carter-Wallace, Inc., used a print advertisement for Pearl Drops toothpaste which claimed that "Laboratory Testing Proves: Pearl Drops Cleans Better Than Rembrandt For Half The Price," "In Laboratory Testing, Pearl Drops Outcleaned Rembrandt by 35%," and "To Help Get Your Teeth Their Whitest Use Pearl Drops Whitening Toothpaste" (82). The Den-Mat Corporation, the manufacturer of Rembrandt Whitening Toothpaste, challenged these claims. Prior to NAD's investigation, Carter-Wallace withdrew the claim that Pearl Drops outcleaned Rembrandt by 35%. In support of the other claims, Carter-Wallace cited FTC and federal cases upholding comparative claims where the claims were based truthfully on laboratory studies and described laboratory tests it relied upon that determined the cleaning abilities of various dental products. Based on this information, NAD found that the claims were substantiated. The NAD noted further that the challenger had failed to present consumer research substantiating *its* position (83). This case reflects NAD's view that a superiority claim may properly be made in an advertisement when the source of support for the claim is clearly disclosed and provides a reasonable basis for the claim.

Finally, a September 1995 NAD decision provides some insight into the degree of substantiation required in NAD's view to support a comparative superiority claim. Farouk Systems, Inc., claimed in its print advertisements for various of its hair-coloring and perming products that the products had a 12-year shelf life and were the "Best," "Longest Lasting," and "Fastest" products to use. Specifically, the print ads contained statements such as "100% guaranteed to be the best perm you've used!" and "100% guaranteed to be the fastest, safest, simplest and longest lasting color in the world" (84). Clairol, Inc., another prominent manufacturer of hair-coloring and perming products, brought these advertisements to NAD's attention and challenged the basis for such claims. Although Farouk presented comparison-test data in support of its claims, the comparison tests were conducted with only a select few of the other leading hair-color and perm products. NAD concluded:

> Overall superiority claims require substantiation that reflects as much of the entire market as can practically be considered, not just the leaders even if the claim might be accurate vis à vis that portion of the market. In this instance, to support a "best perm" or "longest lasting color," NAD agrees with the challenger in that every shade and perm from Farouk would have to be tested against every shade and perm from the competitors according to a test that was well designed and executed. NAD also acknowledges that the challenger has shown that Farouk color services can take as long as one of its competitors (85).

Farouk subsequently agreed to revise its claims.

NETWORK AD CLEARANCE DIVISIONS

Yet another source of control over cosmetic claims are the major broadcast television networks. Television networks do not usually judge local advertisements. However, they do review national advertising campaigns. In some cases, television networks will refuse to run a company's ad unless the company provides adequate substantiation for its product claims.

For example, the American Broadcasting Company's (ABC's) Department of Broadcast Standards and Practices policy (86) states several principles designed primarily for comparative advertising but which are applicable to all advertising formats. One requirement is that comparative claims must be based on testing and surveys "conducted in accord with generally accepted scientific and technical procedures" that are adequate for comparison purposes. In addition, the advertiser must establish that it has selected the "best possible" tests as proof of superiority and must clearly disclose the limitations of such studies. Furthermore, conclusions based on test results must be based on product characteristics that are meaningful to the typical consumer of the product. Finally,

> Regardless of technical compliance with the foregoing PRINCIPLES, if, in the judgment of the Department of Broadcast Standards and

Practices, the *net impression* of the commercial announcement is misleading, deceptive, vague, equivocal or disparaging, it shall be deemed unacceptable for broadcast (87).

The National Broadcasting Company (NBC) enforces similar rules with respect to comparative advertising. First, competitors must be "fairly and properly identified." In addition, advertisers may not attack competitors or their products, services, or other industries by using direct or implied claims that are "false, deceptive, misleading or have the tendency to mislead." The advertising must compare significant properties and should contrast related ingredients, dimensions and features "wherever possible, by a side-by-side demonstration." For all advertising, NBC can require substantiation to be updated and may review supporting data where necessary in the event of a challenge (88).

CONCLUSION

Manufacturers of cosmetic products should ensure that each objective claim about their products, whether on the label, in other packaging and labeling, or in advertising, is appropriately substantiated. As the above review demonstrates, there is no single regulation or formula for determining precisely what amount and type of substantiation is adequate for a particular product and claim. Instead, companies should use common sense and good judgment in deciding what sources to turn to for data supporting cosmetic claims. Companies have relied on animal studies, human studies, consumer preference and comparison testing, as well as published scientific literature and the opinions of hired experts. A variety of substantiation methods are reviewed in other chapters.

REFERENCES

1. Title 21, United States Code (U.S.C.) § 301 et seq.
2. 15 U.S.C. § 1451 et seq.
3. 15 U.S.C. § 41 et seq.
4. The authority vested in the Secretary of the DHHS by the FDC Act and the FPLA has been delegated to the Commissioner of Food and Drugs, who directs the FDA. Title 21, Code of Federal Regulations (C.F.R.) § 5.10 (a).

5. FDA Compliance Policy Guides (CPGs), Ch. 55m, Guide No. 7155m.01. *See also* Working Agreement Between FTC and FDA, 4 Trade Reg. Rep. (CCH) ¶ 9,850.01 (1971).
6. CPG Guide No. 7511m.01 at 2–3.
7. 21 U.S.C. § 321 (i).
8. 15 U.S.C. §§ 1454, 1456, 1459 (a).
9. *See* 21 U.S.C. § 361 (a) and corresponding FDA regulations in 21 C.F.R. Part 700.
10. *See* 21 U.S.C. §§ 321 (t), 361 (e), 362 (e), 379e and corresponding FDA regulations in 21 C.F.R. Parts 73, 74, 81, 82.
11. *See* 21 C.F.R. § 740.10 (a).
12. *See* 15 U.S.C. § 1453 (a), 21 U.S.C. § 362 (b), and FDA regulations in 21 C.F.R. Part 701.
13. 21 U.S.C. § 362 (a).
14. 21 U.S.C. § 321 (n).
15. FDA Regulatory Procedures Manual (RPM), 8-10-10.
16. *Id.* at 8-10-70.
17. 21 C.F.R. § 7.45 (a) (1).
18. 21 C.F.R. § 7.40 (b).
19. 21 C.F.R. §§ 7.45 (c), 7.46.
20. 21 U.S.C. § 334.
21. 21 U.S.C. § 332.
22. 21 U.S.C. §§ 331, 333, 335.
23. 15 U.S.C. § 1456.
24. DHHS, Final Report of the Advisory Committee on the Food and Drug Administration, May 1991.
25. John E. Bailey. Skin Care— State of the Art; A Regulatory View— Alpha-Hydroxy Acid. Presentation to the New York Chapter, Society of Cosmetic Chemists Annual Spring Seminar, April 6, 1994.
26. DHHS, Justification of Estimates for Appropriation Committees, January 1, 1994 (hereinafter, "Estimates 1994"). This figure for food establishment inspections does not include the additional 7,794 inspections of food production facilities conducted pursuant to joint federal/state contract.
27. DHHS, Justification of Estimates for Appropriation Committees, January 1, 1995 (hereinafter "Estimates 1995"). Again, this figure for food establishment inspections does not include the additional 7354 inspections of food manufacturing facilities conducted pursuant to joint federal/state contract.
28. Estimates 1994, Estimates 1995.
29. Estimates 1994.
30. Estimates 1995.

31. 21 U.S.C. § 321 (g) (1).
32. 409 F.2d 734 (2d Cir. 1969).
33. *Id.* at 737.
34. 415 F.2d 369 (3d Cir. 1969).
35. *Id.* at 371.
36. 21 C.F.R. Part 207
37. 21 C.F.R. Parts 210, 211.
38. 21 U.S.C. § 321 (p). *See also* FDA regulations pertaining to drug approval applications in 21 C.F.R. Part 314.
39. 15 U.S.C. § 55 (e).
40. 15 U.S.C. § 45 (a) (1).
41. *See, e.g., Cliffdale Assoc., Inc.*, 103 F.T.C. 110, 163 (1984) and the FTC Deception Policy Statement appended thereto.
42. 15 U.S.C. § 52.
43. 15 U.S.C. § 52 (a) (1).
44. 15 U.S.C. § 52 (b).
45. *See, e.g., F.T.C. v. Simeon Management Corp.*, 532 F.2d 708 (9th Cir. 1976).
46. 15 U.S.C. § 55 (a) (1).
47. *See* 21 U.S.C. § 321 (n).
48. *See* 15 U.S.C. § 57b-1 (statutory requirements for issuing civil Investigative Demands).
49. 15 U.S.C. § 45 (b).
50. 16 C.F.R. § 2.31.
51. 16 C.F.R. § 2.32.
52. 16 C.F.R. § 2.34.
53. 16 C.F.R. § 2.41.
54. The FTC may also bring an action in a federal district court. 15 U.S.C. § 56 (a) (2).
55. For example, upon determining that Listerine was not an effective cold remedy, the FTC prohibited the Warner-Lambert Company from advertising its mouthwash for a certain period unless the advertisement affirmatively stated that Listerine would not help, ease, or prevent colds or sore throats. *Warner-Lambert Co. v. F.T.C.*, 562 F.2d 749 (D.C. Cir. 1977), *cert. denied*, 435 U.S. 950 (1978).
56. 15 U.S.C. § 57b (b).
57. *See generally* 15 U.S.C. § 45.
58. This enforcement mechanism was first established by the Commission in *Pfizer, Inc.*, 81 F.T.C. 23 (1972).
59. *National Dynamics Corp.*, 82 F.T.C. 488 (1973), *aff'd. and remanded on*

other grounds, 492 F.2d 1333 (2d Cir. 1973), *cert. denied*, 419 U.S. 993 (1974), *reissued* 85 F.T.C. 391 (1976). *See also*, Policy Statement Regarding Advertising Substantiation Program, 49 Fed. Reg. 30999–31001 (Aug. 2, 1984).

60. Comments of the Bureaus of Competition, Consumer Protection and Economics of the Federal Trade Commission to FDA Docket No. 85N-0061 in response to request for comments on proposed FDA regulations regarding public health messages on food labels and labeling (1987) (hereinafter, "FTC Health Claims Comments") at 13.

61. *See Pfizer, Inc.*, 81 F.T.C. 23, 64 (1972); *Thompson Medical Co., Inc.*, 104 F.T.C. 648, 813, 821 (1984), *aff'd*, 791 F.2d 189 (D.C. Cir. 1986), *cert. denied*, 479 U.S. 1086 (1987); *Bristol Myers v. FTC*, 102 F.T.C. 21, 321 (1983), *aff'd*, 738 F.2d 554 (2d Cir. 1984), *cert. denied*, 469 U.S. 1189 (1985).

   A 1987 FTC consent agreement with a toothpaste manufacturer demonstrates the range of what the FTC considers a "reasonable basis" sufficient to substantiate a product claim. *In re Jerome Milton, Inc.*, Trade Reg. Rep. (CCH) ¶ 22,468 (1987). For claims relating to hot and cold sensitivity and alleviation of canker sores, FTC required at least one double-blind, well-controlled clinical trial. For claims relating to plaque reduction and gingivitis, the FTC required at least two such trials. For other general claims, the FTC required objective evidence obtained through tests, studies, research and analysis conducted by qualified persons.

62. *Pharmtech Research, Inc.*, 103 F.T.C. 448, 459 (1984). *See also, Viobin Corp.*, 108 F.T.C. 385, 394 (1986).

63. FTC Health Claims Comments at 13–14.

64. *See, e.g., AHC Pharmacal, Inc.*, 101 F.T.C. 40 (1983).

65. *See, e.g., American Home Products Corp.*, 98 F.T.C. 136, 368 (1981).

66. *Bristol Myers Co.*, 46 F.T.C. 162, 175 (1949), *aff'd on other grounds*, 185 F.2d 58 (4th Cir. 1950). ("Concerning the representation that Ipana toothpaste will beautify the smile and brighten and whiten the teeth, the Commission is of the opinion that the reference to beautification of the smile was mere puffery, unlikely, because of its generality and widely variant meanings, to deceive anyone factually.")

67. *See, e.g., Ward Laboratories, Inc. v. F.T.C.*, 276 F.2d 952 (2d Cir. 1960), *cert. denied*, 364 U.S. 827; *Wybrant System Products Corp. v. F.T.C.*, 266 F.2d 571 (2d Cir. 1959), *cert. denied*, 361 U.S. 883.

68. *See, e.g., Gelb v. F.T.C.*, 144 F.2d 580 (2d Cir. 1944); *Herbold Lab, Inc.*, 47 F.T.C. 1304 (1951).

69. *See, e.g.*, *Hall & Ruckel, Inc.*, 32 F.T.C. 229 (1940).
70. As previously described, part of the FDA's enforcement effort in this area consisted of a barrage of warning letters to companies making "drug"-type claims that their products delayed or reversed the effects of aging.
71. *Revlon, Inc.*, FTC Docket No. D09231, amended complaint, 2 (Sept. 7, 1989).
72. *FTC v. Silueta Distributors, Inc.*, 1995-1 Trade Cases (CCH) ¶ 70,918 (N.D. Cal. Feb. 24, 1995).
73. Council of Better Business Bureaus, Inc., "Dear *** , Your Advertising Has Recently Come to the Attention of the National Advertising Division," Guide for Advertisers and Advertising Agencies, 1983 (hereinafter "NAD Guideline").
74. NAD Guideline at 1.
75. *Id.*
76. *Id.*
77. *Id.* at 3.
78. 24 (6) NAD Case Reports 91 (Aug. 1996).
79. *Id.* at 91–92.
80. 24 (4) NAD Case Reports 55 (June 1994).
81. *Id.* at 56.
82. 23 (9) NAD Case Reports 95 (Dec. 1993).
83. *Id.* at 96.
84. 25 (7) NAD Case Reports 182 (Sept. 1995).
85. *Id.* at 188.
86. ABC, Dep't of Broadcast Standards and Practices, *Principles for Comparative Advertising* 1.
87. *Id.* at 2.
88. NBC, *Comparative Advertising Guidelines—NBC Television Network*.

2
Views on Claims Support Methods for Hair Care Products

Elaine Stern, Al Gabbianelli, Amy Qualls, Judith Rahn, Grace Yang, and Norman Meltzer
Helene Curtis, Rolling Meadows, Illinois

INTRODUCTION

Claims are one of the principal ways in which consumers are influenced to try outstanding new products. They inform and educate the consumer about what a hair care product will do and why that product works. Because claims play such an important role, marketers have an obligation to efficiently and effectively get clear and truthful messages about their products to the consumer.

The way claims are structured and supported reflects on corporate values and the corporation's philosophy about consumers. For this reason, advertising copy and support documentation should always be checked by designated senior executives for consistency with corporate policy. In addition to this internal review, outside forces have a very significant influence on the nature of advertising claims. Television networks examine advertising copy for compliance with their standards before an ad is aired, and it is not uncommon for an advertiser to make changes to claims in response to the networks. Also, if advertising copy is challenged by competitors or consumer protection groups, television networks, or the National Advertising Division of the Council of Better Business Bureaus, then the National Advertising Review Board, the

Federal Trade Commission, and the courts can become involved. Successful advertisers maintain a current awareness of how cases that come before these agencies are settled and use this information in developing a strategy for claim support.

During the past 30 years, the physical, chemical, and optical techniques available for supporting hair care claims have become more sophisticated as our understanding of physiology, hair structure, and the physical properties of hair have advanced dramatically. Yet, the basic principle of claims support, getting truthful messages to the consumer, has not changed. This requires product developers to have a thorough understanding of the types of claims that can be made and the kind of testing needed to support them. Claims can be either objective or subjective. Objective claims are statements of facts about a product that are generally known or can be proved by using well-designed tests. For objective claims, prior substantiation must exist before they can be made. On the other hand, subjective claims are statements about intangible qualities of a product that can neither be proven nor disproved. If a claim is truly subjective, there are no methods or means by which it can be substantiated.

Objective Claims

Objective claims can be descriptive, implied, comparative, preemptive, or testimonial.

Descriptive Claims

Descriptive claims tell how the product works or what it does. It can be as simple as "cleans and conditions your hair," which requires only minimal claim substantiation, such as laboratory testing on hair swatches or testing in a salon. Descriptive claims can also be complex, such as "works by absorbing protein into each hair fiber," which can require complicated laboratory experiments using radiolabeling techniques.

Implied Claims

Implied claims are not explicit statements about product performance but usually require testing nonetheless. Product performance can be implied by the name of the product or by statements about ingredients used to make the product. A product calling itself YXZ Baby Shampoo must be tested to document its suitability for use on babies. Implied claims are also made by highlighting ingredients such as proteins, vitamins, herbs, or aloe on a shampoo label. These ingredients imply performance benefits and testing is needed to demonstrate that these benefits are perceived by the consumer.

Comparative Claims

Comparative claims refer directly or indirectly to other brands in the category and can assert either superiority or parity. Other brands do not necessarily have to be named. Because comparative claims either name or imply the name of competitive products, claim substantiation documentation for this type of claim is usually scrutinized by both the networks and by competitors.

Superiority Claims. Superiority claims state that a product performs better than another product or a category as a whole. Examples of superiority claims are "Cleans and conditions your hair better than brand X" or "Keeps your hair cleaner longer." This second superiority claim implies that the shampoo cleans better than all other products in the shampoo category. Superiority claims and puffery claims are sometimes difficult to distinguish. Subtle changes in wording will significantly change the intention of the claim and determine whether supporting testing will be required.

Parity Claims. Parity claims are also comparative. These claims state or imply that a product performs as well as another product. An example of a parity claim is "Cleans and conditions your hair as well as brand X, but costs less" or "No other product conditions better than brand X."

Preemptive Claims

Preemptive claims are factual, descriptive claims that are true for most products of the same type but are made in a way to imply uniqueness because no one else has made that claim. A claim such as "Conditions your hair only where you need it" is true for all conditioners. Support for this claim can be done through the use of published literature or through logical argument.

Testimonial Claims

Testimonial claims are statements of an individual's opinion about a product. An example is "I just tried shampoo X and it leaves my hair in better condition than shampoo Y." While a testimonial claim may literally be true for the particular person making the statement, testimonial claims of opinion must be supported by data demonstrating that the opinion is projectable to all consumers. Testimonial claims would require consumer preference testing.

Subjective Claims

Subjective claims are statements considered puffery and therefore cannot be proved. Puffery claims are laudatory statements about a product that are so vague, general, or overly enthusiastic that consumers recognize and accept them as statements that need not be proved. Examples of these are "For salon beautiful hair," "Makes you feel beautiful all day long," and "The best you can get." Sometimes, the line between puffery and an objective claim requiring claim substantiation is not easy to distinguish, and subtlety of wording can make the difference.

Advertisers of hair care products are developing ever more assertive and daring claims. As new brands are launched and existing brands are repositioned, exciting, aggressive claims grab the attention of consumers and alert them to new product offerings and what to expect from them. At the same time, the advertiser is faced with the challenge of meticulously documenting product performance, so that the claims made truly reflect the product's attributes.

This chapter describes the testing methods that are being used in developing state-of-art hair care products and substantiating descriptive and comparative claims about them. Protocols that involve instrumental, clinical, and sensory techniques will be discussed. The key feature of each of these methods is that they have been successful in predicting consumer perception of product performance. In the final analysis, development of a successful hair care formulation and claim support program depends solely on how well the evaluation techniques provide an understanding of how the consumer will respond to a new product offering. Case histories will give examples of how these kinds of methods have been used and misused over the years in developing and marketing hair care products.

INSTRUMENTAL METHODS

A wide range of instrumental techniques are routinely used to provide support for substantiating hair care claims. For example, tensile measurements are commonly used to measure changes in the hair's physical properties and combing attributes. Alterations in surface characteristics and hair damage are can be assessed by light and electron microscopic analysis. Energy dispersive x-ray analysis (EDX) is one of the many techniques used to confirm and investigate the deposition of conditioners based on silicone polymer technology. In a corresponding manner, the deposition, substantivity, and penetration of organic materials such as permanent wave actives, protein hydrolysates, and vitamins are determined through radio chemical labeling and analysis. Subtle changes in the hair's shine and luster have been measured by goniophotometry and, more recently, by color image analysis (IA). Complex changes in the hair's flexibility and rigidity can now be evaluated by dynamic mechanical analysis (DMA).

This chapter summarizes salient instrumental techniques and applications that are used to support some of the most important hair care claims. A list of typical products, the hair care attributes which they target and the applicable evaluation techniques for each is presented in Table 1.

Physical Properties of Hair

Assessment of the hair's physical properties and attributes can be divided into two major categories: single fiber techniques and collective assessment techniques (1). Primary assessment techniques measure the basic physical and tensile characteristics of the individual hair fibers. The cumulative contributions of these characteristics give rise to the more familiar "collective" hair attributes such as ease of combing, shine, static fly-away etc. These gross attributes are readily evaluated by collective assessment techniques. Secondary techniques generally use

Table 1 Routine Claims Substantiation Procedures for Shampoos, Conditioners Styling Aids, Hair Sprays, and Permanent Waves

Product	Claim	Test method
Styling aids	Set hold	Dynamic and static set Retention
Styling aids	Curl durability, strength	Curl compression
Styling aids	Increased volume / body	Radial compression
Permanent waves	Perm efficiency	Tensile kinetics and stress Relaxation studies
Shampoos, conditioners, two-in-ones	Cationic conditioner / buildup claims	"Rubine" dye absorption
Shampoos, conditioners, two-in-ones	Silicone conditioner / buildup claims	Mid-FTIR, EDX
Shampoos, conditioners two-in-ones	Static buildup, ballooning, fly-away	Triboelectric measurement
Shampoos, conditioners, two-in-ones	Luster and shine	Goniophotometry, color image analysis
Shampoos, conditioners, two-in-ones, chemical treatments	Ease of combing and detangling	Dry and wet combing
Shampoos, conditioners, two-in-ones, chemical treatments	Hair smoothness, improved cuticle condition	Dry and wet friction, microscopic analysis
Shampoos, conditioners, two-in-ones, chemical treatments	Hair strength, resiliency, and elasticity	Dry and wet tensile strength

hair tresses to measure gross changes that occur as a result of product treatment.

Primary Assessment Techniques

Tensile Properties

Possibly, the most important attributes that can be assessed by single fiber techniques are tensile properties. Standardized tensile testing procedures are routinely used to measure hair strength (break stress), longitudinal elasticity (yield stress), and stiffness (modulus). Virtually all tensile testing procedures deal with stress, strain, elastic modulus, yield, and break points. A complete list of mechanical testing definitions has been compiled by Busche (2).

Normally, measurements of modulus, yield stress; and break stress are taken on control and treated samples in order to evaluate product-induced changes and substantiate claims surrounding the hair's physical attributes. Modulus provides information on the internal stiffness or softness of the fiber, yield stress provides data on hair elasticity, and break stress is used to assess changes in the actual strength of the fiber. Treatments such as bleaching, chemical straightening, oxidative dyeing, and permanent waving have a significant impact on tensile properties.

Intermittent Stress Relaxation Techniques

Other tensile procedures such as stress-relaxation studies have classically been used to evaluate the efficiency of permanent waves and to provide support for claims regarding the behavior of commercial waves. Szadurski and Erlemann (3,4) originally developed the hair loop test as a means to determine the softening effect, aggressivity, efficiency, and optimal processing time of permanent waves. Newer techniques such as single fiber tensile kinetics (5,6) utilize a more sophisticated approach to assess the behavior and efficiency of permanent waves. While the newer techniques have significantly expanded our understanding of the permanent waving process, their use in the claims support arena has been limited.

During permanent waving, disulfide cross links are broken and then reformed in order to alter the hair's structural properties. These stress-relaxation techniques indirectly estimate the rate and extent of bond breakage and reformation by measuring subtle tensile changes as a hair fiber is stressed and relaxed during a simulated permanent wave. More efficient wave lotions produce quicker decreases in stress, while more efficient neutralization, conversely, produces rapid increases in stress. Increased perm efficiency and neutralization broadly support claims for more effective perms that produce more resilient curls.

Friction

As a comb passes through the hair, varying levels of friction are produced at the comb/hair-fiber interface. The ensuing frictional drag opposes the comb's progress and contributes to the hair's combing characteristics. Resistance to combing is, therefore, attributable in part to each hair fiber's contribution to the total friction. Products such as stand-alone shampoos and conditioners and two-in-one conditioning shampoos enhance combability by minimizing friction at the fiber surface. Hence, the assessment of product-induced changes in frictional attributes is a critical step in supporting claims for product performance.

The most popular methods for assessing kinetic friction are based on modifications of the techniques presented by Schwartz and Knowles (7) and Scott and Robbins (8). These techniques utilize a single rotating capstan in conjunction with a tensile tester to measure the coefficient of kinetic friction (μk). Either a single hair fiber or a bundle of hair fibers is wrapped once around the mandrel. The root end of the single fiber or bundle is connected to the load cell of a tensile tester, while a standard weight (1 to 10 g) is suspended from the tip end. Product- or treatment-induced changes in the frictional properties of the hair are then determined.

Collective Techniques

The combability of human hair is predominantly derived from the attributes of individual hair fibers. Factors such as diameter, surface roughness, friction, static charge, length, and fiber-to-fiber interactions (i.e., tangling and knotting) contribute to combability characteristics (9–12). Hair treatments such as bleaching, chemical straightening, oxidative dyeing, and permanent waving usually have a deleterious effect on fiber properties and increase combing difficulty. Conversely, shampoos, conditioners, conditioning (two-in-one) shampoos, and other conditioning treatments improve combing attributes by ameliorating the negative effects associated with hair damage. Claims on improved combability and decreased tangling are directly supported by standard instrumental combing procedures.

Combing Measurements

The most popular techniques for measuring combability stem from the work of Garcia and Diaz (13). Their basic technique utilized a tensile tester to measure the force required to pull a hair tress through a comb at a constant speed. As the hair is pulled through the comb, measurements of peak load, average load, and energy are used to assess overall combability. Peak load values reflect the degree of tangling that the comb encounters as it passes through the tress; higher peak force values indicate greater tangling. Average load readings are taken as the comb traverses the tress from the hair root to tip. Average load is indicative of the hair's physical state. Hair damage such as uplifted cuticles and split or fragmented fibers requires greater combing force, which increases combing difficulty. Energy is a measure of the total work expended in combing the hair tress.

Combing measurements are performed on dry and wet hair to simulate daily combing and "after washing" conditions respectively. Wet combing measurements are made immediately after product treatment, while dry combing measurements are routinely taken after equilibrating the hair tresses at a specific temperature and humidity.

Cationic and Silicone Deposition

Cationic surfactants such as centrimonium chloride are commonly used as hair conditioners. These materials substantially improve overall condition by ameliorating the damaging effects produced by bleaching, oxidative dyeing, permanent waving, blow dryers, curling irons, and natural weathering effects. Unfortunately, exaggerated and/or prolonged use of these materials results in buildup, which creates heavy, limp hair that is not amenable to curling, permanent waving, and other "bodifying" treatments. The determination of conditioner deposition and buildup is therefore of critical importance in the evaluating cationic based conditioners and in providing support for conditioning and "no buildup" claims.

Scott et al. (14,15) pioneered the use of dyes to assess the sorption/desorption profiles of cationic surfactants on hair fibers. They found that hair fibers that had adsorbed cationic conditioners would stain red when exposed to an appropriate dye, the intensity of the stain being roughly proportional to the adsorbent's concentration. In the ensuing years, a multitude of procedural modifications and alternative dyes have been recommended (16,17). Current procedures (18,19) entail washing natural white hair tresses or wool swatches with a suitable cleanser. The hair tresses or swatches are thoroughly rinsed, treated with the cationic conditioner, and rigorously rinsed. Multiple washings and applications are used to simulate "buildup" conditions. The treated tresses are exposed to a suitable dye, rinsed, and dried. Conditioner deposition and/or buildup is demonstrated when the test tresses adsorb the dye and change color. The test provides dramatic visual evidence of conditioner deposition, which is particularly useful in demonstrating and substantiating deposition claims. Buildup is demonstrated by a proportional change in the intensity of the color response. In a similar manner, mid-infrared spectroscopy can be used to qualitatively assess silicone deposition and buildup on tresses after treatment with two-in-one conditioning shampoos.

Luster or Sheen

One of the primary goals of hair care technology has been to develop products that provide and/or enhance the natural sheen and luster associated with healthy hair. Consequently, a variety of techniques have been developed to evaluate products and substantiate claims related to hair luster. Some of the earliest measurements of hair luster were performed by Thompson and Mills (20) using a device that measured the reflectance of aligned hair fibers that were rotated within a light beam to give maximum reflectance. Den Beste and Moyer also describe a technique (21) that involves the measurement of contrast ratios. Stamm et al. (22) further developed the technology by using a goniophotometer to measure the reflectance of polarized light from a parallel array of taut hair fibers. A more complete description of these techniques is provided by Robbins (23).

Set Retention

Styling aids, hair sprays, and permanent waves seek to maintain the hair in an attractive and pleasing configuration. The effects of these products range from short term for styling aids and hair sprays to long term for permanent waves. Each product is therefore designed to be efficacious within a given time frame. Unfortunately, changes in humidity, temperature, and daily combing habits can adversely affect the set-holding properties of styling aids and hair sprays. In a corresponding manner, permanent waves can be affected by excessive shampooing, conditioning, bleaching, and dyeing. Hence, a number of test procedures have been specifically designed to assess the efficacy and durability of these hair treatments (24–29). The majority of these methods assess either dynamic or static set-retention properties.

Dynamic set retention is used to assess the set-holding properties of hair sprays and other styling aids. In this procedure, hair tresses are treated with the test product, mounted on a setting jig in a zig-zag configuration, and equilibrated under standard conditions. The tresses are removed from the jig and mounted on an enclosed tensile tester with the root and tip ends respectively secured to the tester's upper and lower clamps. Each tress is then stretched (under constant physical strain) until fully extended and allowed to recover. This stretching and recovery cycle is repeated in triplicate at different time intervals (0.5, 2, and 4 h). The resultant data express the dynamic set retention as a function of time and can be used to support claims related to inherent set-holding properties.

Curl Strength

A styling aid's ability to impart curl strength can also be determined from the load-extension values obtained during the dynamic set-retention tests. Load extension data provides a measure of the amount of force required to extend (or uncurl) a curl to its straightened position. Higher peak load values are indicative of a stronger set and tighter curl formation.

Resiliency and Bounce

Claims for the resiliency and/or bounce imparted by a hair treatment can be supported from data generated during set-retention tests. Resiliency is simply defined as the ratio of the length of the set hair before the application of strain at time 0 divided by the length of set hair after application of strain at time t.

Static Set Retention or Curl Retention

The static set-retention or curl-retention test (25,30) is performed in much the same manner as the dynamic set-retention test. The major procedural difference in this test is that tresses can be set on standard circular rollers (or multipin jigs). Curl fatigue is achieved via natural gravitational force rather than by applied physical strain. Treated tresses are mounted on rollers or pegboards and equilibrated under standard conditions. The conditioned tresses are then removed from the rollers and hung in a controlled-environment incubator. Static set-retention measurements are made at timed intervals as the curls relax due to gravitational force. These types of measurements are appropriate for supporting claims pertaining to the duration of set.

Static Charge

As a comb passes through the hair, electrical charges develop on the comb and the hair fibers. Since neither of these materials can act as a conductor to effectively dissipate the electrical buildup, a "static" charge develops. Ballooning or fly-away occurs when neighboring hairs repel each other as charges of the same sign and magnitude accumulate on the fibers. Consequently, the hair becomes difficult to comb, unmanageable, and damaged if the condition is allowed to persist. One of the primary goals of conditioning agents is to improve combing and styling attributes by alleviating static buildup. Numerous techniques (31–37) have been developed to evaluate the antistatic performance of shampoos and conditioners and in turn to substantiate claims for controlling hair static and fly-away. Static buildup can easily be determined by measuring the electrical charge generated by a comb as it passes through a dry hair tress. After a thorough cleaning, the tresses are treated with the test materials and brought to uniform dryness by equilibration in a controlled environment for a prescribed time. Typically, a tress is mounted on a combing jig that is equipped with an electrostatic voltmeter and contained within a Faraday cage. The geometry of the voltmeter and tress positions are standardized to optimize the static readings and generate reproducible results. Combing of the tress should be automated so that the number of strokes, rate of combing, and applied pressure remain constant.

Radio Assay Techniques

A variety of techniques have been to support claims regarding the penetration of cosmetic actives (e.g., vitamins, proteins and amino acids) into the hair fiber. Classical radioassay techniques provide the most direct and reliable method for substantiating claims on the adsorption and penetration characteristics of hair care actives (38–41).

When hair is treated with a cosmetic active, the material may either rinse off, adsorb to the hair surface, and/or migrate to the interior of the fiber. If the active has been labeled with either ^{3}H or ^{14}C, the amount of surface deposition can easily be determined by measuring the radiation on the exterior of the intact fibers. Since radiation emitted by material that has diffused into the fiber cannot be detected at the surface, the hair can subsequently be digested and analyzed for residual radioactivity. The difference in radioactive content between these analyses represents the material that has penetrated into the hair. In a similar manner, the amount of material that has diffused into the cuticle and cortex can be determined by separating the cuticle from the cortex and analyzing each fraction independently.

Microscopy Techniques

Microscopy provides a versatile set of tools useful for both the substantiation of claims and the illustration of claims substantiated through other means. Essentially any claim that can be translated into a "visible" modification of the surface appearance or physical condition of the hair fibers can be addressed using microscopy techniques. Microscopy results may be quantitative, semiquantitative, or qualitative (descriptive) in nature. They are uniquely valuable, however, when presented in pictorial form, as *photomicrographs*. Photomicrographic evidence allows the audience to "see" the attribute being claimed. In the case of side-by-side comparisons (before versus after, treated versus untreated, etc.) a photomicrographic presentation can be especially persuasive and dramatic.

Scanning Electron Microscopy

Scanning electron microscopy (SEM)* facilitates detailed examination of the hair fiber surface at much higher magnifications and with a much greater depth of focus than is possible using an light microscope. Cuticle appearance, surface debris, and thick surface deposits are readily observed; hence, SEM results are frequently applied to claims of cleansing, damage prevention, or damage reduction. Many examples where the SEM has been used in the evaluation and illustration of hair care product performance can be found in published literature (42–51).

When using traditional instrumentation, hair fibers (and any applied product) must be dry before SEM examination. As SEM uses electrons instead of light to generate images, it provides no information regarding such properties as color and gloss.

Claim Support Applications.   Claim-related information obtained using SEM commonly falls into one of two categories: information regarding cuticle irregularities (damage) and information regarding surface deposits (styling aids, debris, etc.).

Cuticle Irregularities (Damage/Conditioning).   Treatment effects on the physical condition of the hair cuticle (both beneficial and deleterious) are routinely studied by SEM. Gross changes may be observed by examining and comparing a suitable number of treated fibers to untreated controls. Where subtle changes are of interest, it may be preferable to examine the same fiber before and after treatment. Direct before and after examination, however, is not possible when the fibers have been conductively coated prior to examination.

Some examples of the types of features and changes that may be observed are illustrated in Figures 1–5. Figures 1 and 2 demonstrate the effect of a conditioner treatment on uplifted cuticle. The cuticle scales appear much smoother on the conditioned fiber (Fig. 2) and lie flat against the hair shaft. Note that the conditioning agent itself cannot be observed on the hair fibers, as very thin, fairly uniform coatings are transparent in the SEM. The related claim of damage *prevention* might

---

*Readers unfamiliar with scanning electron microscopy are encouraged to consult Goldstein et al. (52) or Gabriel (53) for additional information regarding its instrumentation, technical requirements, and techniques.

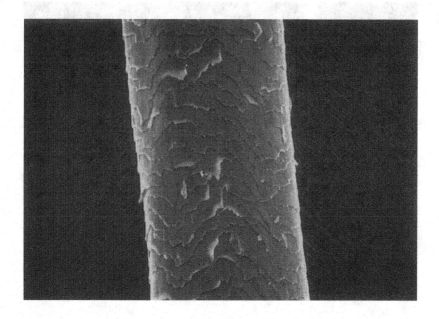

Figure 1   Untreated damaged hair fiber.

be supported by a similar examination of conditioned and untreated fibers after combing or some other mechanical process.

Frequently, the SEM is used to screen chemically treated hair (e.g., permed, bleached, or dyed) for surface damage. Figure 3 shows a fiber treated with a "low damage" type of waving solution. The fiber exhibits no obvious treatment-related cuticle damage. For the purpose of illustration within this document, the fiber in Figure 4 was treated with an intentionally severe wave formulation. The cuticle has obviously been mechanically and/or chemically damaged during treatment.

Surface Deposits (Coatings, Debris, and Removal).   Although very thin conforming films are frequently not evident when examined using SEM, thicker films from styling aids (gels, hair sprays, and occasionally mousses) can usually be observed. Coverage, manner of adhesion (spot weld versus seam), and type of failure (adhesive versus cohesive) may be discerned. Examples of some features commonly observed

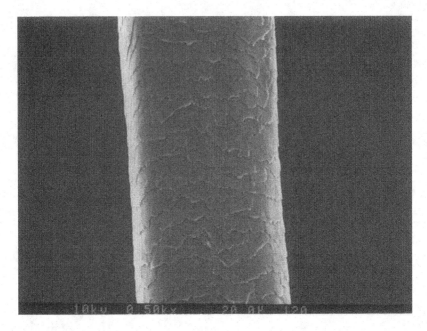

Figure 2   Damaged hair fiber after conditioner treatment.

are given in Figures 5 and 6. In Figure 5, the fibers are covered with a thin, fairly uniform coating of styling aid. To a large extent, the coating conforms to the morphology of the underlying cuticle. As a result, although there is some loss of edge definition, the cuticle has not been completely masked. The existence of the coating becomes obvious by examining the fiber at the left of the photomicrograph, where a portion of the coating has broken away from the fiber.

The fibers in Figure 6 have also been treated with styling aid. Note that in this figure, the deposition is much less uniform, occurring as thicker blotches that frequently obscure the cuticle scales. This type of irregular deposition is much easier to observe than the uniform deposition presented in Figure 5.

Other types of surface deposits, debris, and their removal may also be examined. Figure 7 shows a styling aid–treated hair fiber. The companion photomicrograph (Fig. 8) shows a similarly treated fiber after shampooing. This type of examination might be performed to address issues of cleansing (for a shampoo) or ease of removal (for a styling aid product).

Light Microscopy

Images obtained using light microscopy are typically lower in magnification than SEM images, with less depth of focus. Nonetheless, light microscopy of hair fibers or hair care products can provide valuable information about product performance. Traditional instrumentation is easily linked to video monitors/recorders, image analyzers, or outfitted for fluorescence microscopy. Additionally, examination using light microscopy has the distinct advantage that the same fiber can easily be examined before and after treatment; often the same region in the fiber may be repeatedly observed.

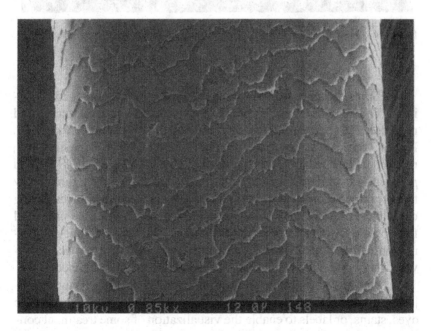

Figure 3 Hair fiber treated with commercial "low damage" permanent wave.

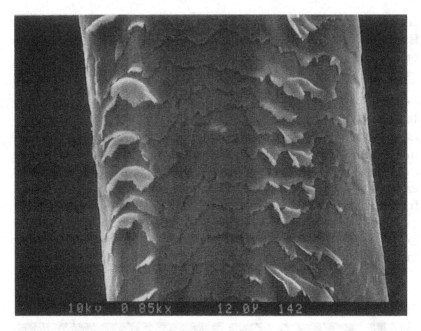

Figure 4   Hair fiber treated with intentionally damaging permanent wave.

Claim Support Applications.   Split-End Prevention and Mending.   The occurrence and mending of split ends are easily observed and recorded with optical microscopy, as shown in Figures 9 through 13. Figure 9 presents a photomicrograph of a split end. The same fiber end, mended using a leave-in type product, is shown in Figure 10. Split-end *prevention* is demonstrated in Figures 11 and 12. These figures show treated and untreated "prone to splitting" fibers at various stages during combing. The formation of a split is clearly observed in the untreated fiber, while the treated fiber remains intact. By tracking split-end formation in a number of fibers, evidence for claims such as "helps prevent" or "reduces the occurrence" of split ends may be obtained.

Dye and Label Studies.   In some cases, hair samples may be subjected to a variety of fluorescent, chromogenic (colored), or metallic dyes, stains, or labels to enable the visualization of some treatment consequence through conventional or fluorescence microscopy. Procedures can often be adopted or adapted from standard histological techniques. Depending on the information sought, either whole fibers or fiber cross sections (sections cut perpendicular to the fiber axis, as shown in Figure 13) may be prepared and examined. When study results are appropriately translated into a specific product claim, compelling, visually attractive evidence for the claim may be obtained.

Several cases in which dyes, stains, or tags have been used to study commercially important attributes are reported. For example, Cooperman and Johnson (54) used a ninhydrin dye to examine the penetration of proteins into hair fibers. Wickett (55) examined the extent of penetration of methylene blue dye into reduced hair fibers on the premise that more reduction would create more porosity and allow deeper penetration of the blue dye into the fibers. Taking a slightly different approach, Evans (56) and Evans et al. (57) used a thiol-specific fluo-

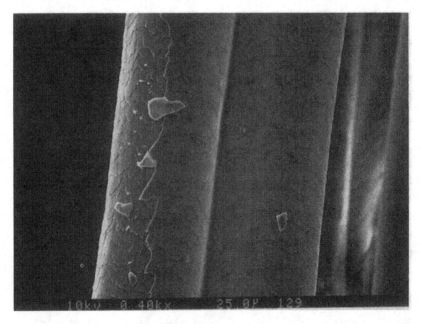

Figure 5 Styling aid–treated hair fiber showing fairly uniform deposition of product.

rochrome to selectively label reduction sites in wool (Evans) and hair (Evans et al.) to gain insight into the extent of reduction and the nature of the reduction reaction.

The substantivity and buildup of cationic polymers and keratin hydrolysate on hair fibers have been investigated by workers at the Textile Research Institute (58) using the fluorochrome uranine. They have also (51) employed both uranine and rhodamine-b to examine hair damage. Jurdana and Leaver (59) used Nile red dye to assist in the characterization of the surfaces of wool and human hair fibers. Many fluorescent probes and their properties are catalogued in Haughland (60). *A Laboratory Manual of Histochemistry* (61) is a collection of standard histochemical techniques and their use presented in "cookbook" style.

Product Discrimination Based on Formulation Attributes.   Occasionally, the inclusion of a specific component in a product formulation

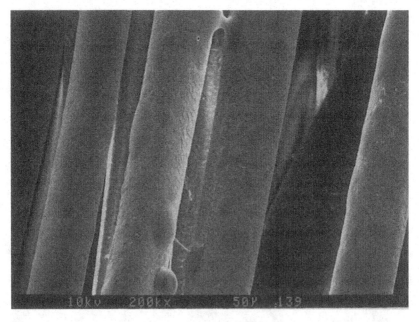

Figure 6   Styling aid–treated hair fiber showing irregular deposition of product.

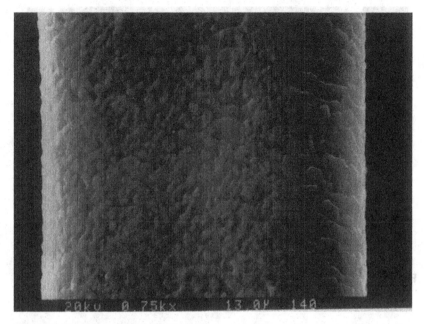

Figure 7 Styling aid–treated hair fiber.

is claimed to provide a particular benefit or to distinguish that product from the competition. When that component is optically or morphologically unique, attractive photomicrographs that illustrate the distinction can be obtained. Perhaps the easiest illustration of this type of comparison pits ordinary shampoo against a shampoo with "conditioning droplets," as shown in Figures 14 and 15.

CLINICAL METHODS

Several clinical testing methods have been developed for the purpose of substantiating product performance claims for hair care products. This section will be devoted to an overview of these various methodologies, specifically in regards to testing which supports those claims commonly associated with shampoos and conditioners marketed to adults, and baby shampoos.

Figure 8 Styling aid–treated hair fiber after shampooing.

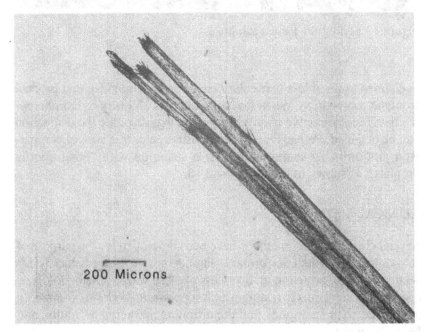

200 Microns

Figure 9 Fiber with "split end."

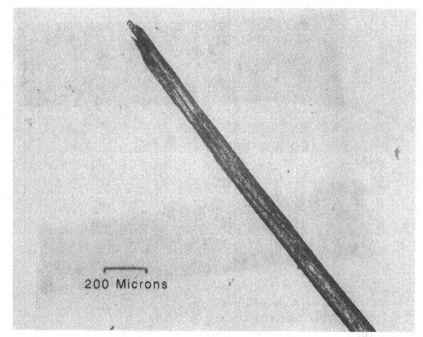

Figure 10   Fiber from Fig. 9 with split mended.

Shampoos

Antidandruff Clinical Evaluations

Popular claims in this category relate to "antidandruff" efficacy. An example of a claim would be "clinically proven to control dandruff." These claims are associated with shampoos containing an active ingredient that has demonstrated antidandruff properties as defined in the OTC monograph (62). Selenium sulfide, sulfur, coal tar, and zinc pyrithione are all active ingredients with established efficacy and are listed in the monograph. A company is not required to independently substantiate the efficacy of an antidandruff shampoo containing one of these materials. However, antidandruff clinicals are necessary in making comparative claims for product performance. Companies also use this clinical method as assurance that the base formulation has not inhibited the efficacy of the active ingredient.

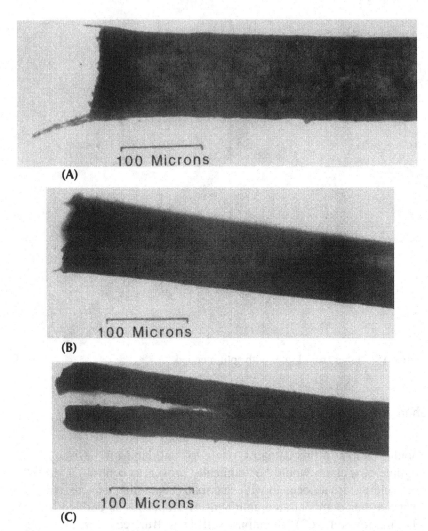

Figure 11 A. Single untreated fiber before combing. B. Single untreated fiber combed 1000 times. C. Single untreated fiber mechanically combed 2000 times.

The main objectives of an antidandruff efficacy clinical are to evaluate a test shampoo on human volunteers displaying moderate to severe dandruff and to assess product safety following use of the test shampoo over a given period of time (usually 8 weeks).

The method to evaluate the efficacy of an antidandruff shampoo (63–72) consists of evaluations of loose and adherent dandruff by a qualified dermatologist on selected panelists who will use the product and are examined at specific time intervals, typically at baseline, 2 weeks, 4 weeks, 6 weeks, and 8 weeks.

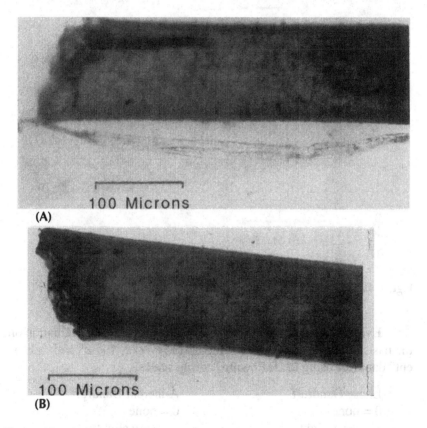

(A)

(B)

Figure 12 A. Single conditioner-treated hair fiber before combing. B. Single conditioner-treated hair fiber combed 1000 times. C. Single conditioner-treated hair fiber combed 2000 times. D. Single conditioner-treated hair fiber combed 3000 times.

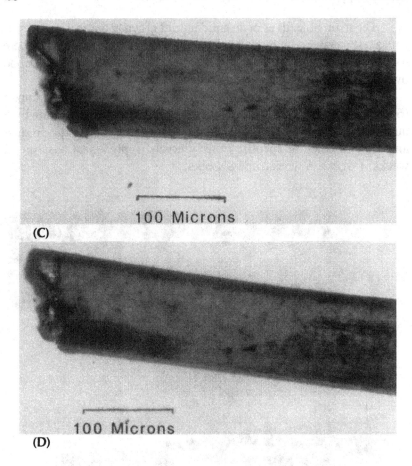

Figure 12 Continued.

Evaluation.   Panelists are qualified for the clinical evaluation on the basis of a dermatologist's rating on the level of "loose" and "adherent" dandruff using the following scoring scales:

Loose Dandruff	Adherent Dandruff
0 = none	0 = none
1 = very slight	1 = very slight
2 = mild	2 = mild
3 = moderate	3 = moderate
4 = severe	4 = severe
5 = very severe	5 = very severe

The severity of both loose and adherent dandruff is determined visually by the examining dermatologist, and a combined loose and adherent score of at least 6 (3 of which is due to adherent dandruff) is required to qualify for a study.

Prospective panelists are excluded if they have any evidence of an active dermatological or scalp condition that could potentially interfere with the integrity of the study.

Test Period. Following the conditioning phase, all the prospective panelists return to the facility for the dermatologist's baseline evaluation. The subjects use test product two times per week according the product use instructions. Other hair care products such as sprays, perms, conditioners, dyes, and shampoos are not to be used by the panelist during the study. Panelists are examined and scored by the dermatologist after 2, 4, 6, and 8 weeks. Each scalp evaluation is performed at least 24 h following the last use of the panelist's assigned shampoo.

Results. The results of a successful antidandruff clinical demon-

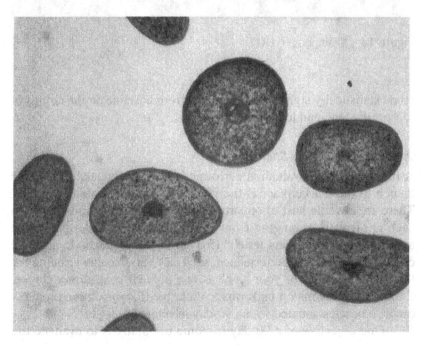

Figure 13 Typical fiber cross section.

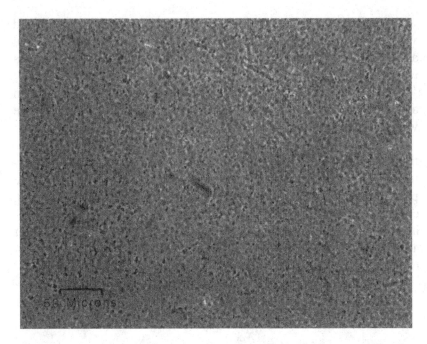

Figure 14   Shampoo product.

strate statistically significant reductions from baseline on the ratings of loose, adherent, and total dandruff scores.

90-Day In-Home Use Studies

A routine clinical evaluation performed on new products prior to the launch to the marketplace is the 90-day in-home use study (73–75). There are a whole host of opportunities for claims substantiation from such clinical trial, ranging from "clinically tested," "dermatologist tested," and "pediatrician tested" (for baby-care products) to attribute claims about product performance, such as "Variable conditioning for the changing needs of your hair," as one popular conditioner claims. Comparative claims on performance relative to competitive products can also be substantiated with a 90-day in-home use study.

One objective of a 90-day in-home use study is to evaluate the safety, efficacy, and performance of the product under semicontrolled in-home use conditions using a population of subjects who typically use a similar marketed product, for example, users of regular shampoo will be enrolled in a 90-day regular shampoo use study instead of users of dandruff shampoo.

The 90-day use studies are performed under Good Clinical Practices guidelines, where there are certain inclusion/exclusion criteria that the subjects must meet prior to being enrolled in a clinical study.

Examples of inclusion requirements are as follows:

Male or female, ages 18 to 65
Individuals in good general health
Individuals who use shampoo products daily
Individuals who are not sensitive to the ingredients of the test
    products, as evidenced by the findings of a 48-h patch test

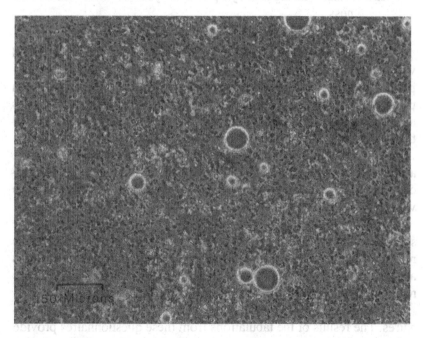

Figure 15 Shampoo product similar to that in Fig. 14 but containing "conditioning droplets."

Individuals who are willing and able to sign an informed consent form and to carry out all procedures of the study

Examples of exclusion requirements are as follows:

Individuals with a history of sensitivity and/or adverse reactions to shampoo products
Individuals with a history of skin allergies or contact dermatitis, especially to cosmetics, fragrances, soaps, and other cleaning agents
Individuals on a regular regimen of steroidal or nonsteroidal anti-inflammatory agents (use of oral contraceptives permitted and noted), antihistamines, or insulin
Individuals with scalp irritation
Females who are pregnant, lactating, or planning pregnancy within the course of the test
Individuals with signs or symptoms of systemic and/or chronic illness

The test procedure consists of several steps.

Subject Selection.   The prospective subjects are patch-tested with the assigned product using the 48-h patch-test protocol. The purpose of this patch test is to determine if any of the test subjects have been presensitized to any ingredients. If there are no significant skin reactions at the patch sites on the back, the scalp of the prospective panelist will then be evaluated for irritation and the hair evaluated for condition.

Test Period.   Subjects are sequentially numbered and issued the assigned study shampoo and diary forms and given instructions concerning usage of the assigned shampoo, study requirements, and return visits.

Return Visits.   Over the course of the test, return visits can be scheduled after 30, 60, and 90 days of test-product usage. Subjects return to the laboratory bringing with them the completed diary forms and the used test product. At each return visit, the diary forms are reviewed and the panelists' scalp and hair are evaluated.

Questionnaires.   At the return visits, panelists fill out questionnaires. The results of the tabulations from these questionnaires provide claims substantiation documentation on product performance attributes. Claim support documentation for a popular two-in-one shampoo claim of a "Self-adjusting formula which gives the right balance between cleansing and conditioning every time" used several attributes from the poststudy questionnaire. The questionnaire was designed to evaluate how active the panelists were in terms of whether they engaged in strenuous, moderate, or low levels of exercise; depending on the type of exercise, hair could become oily or sweaty. The claim "Provides the right balance" was substantiated by panelists rating the product as being highly satisfied on how the two-in-one shampoo did not overclean the hair, providing complete conditioning for dry, brittle hair and thorough, heavy cleansing with light conditioning for oily, sweaty hair.

Overall, the tabulated questionnaires are the primary source of information used in claim substantiation when the results demonstrate statistically significant differences between the attributes or test products under evaluation.

Post patch.   Two weeks after the 90-day evaluation, a second patch test is conducted using the 48-h patch-test protocol. The purpose of a postpatch is to determine whether any of the subjects in the study were sensitized to the test product.

Results.   The final conclusions of a successful 90-day in-home use study demonstrate no clinically significant product-related adverse reactions to the test products.

Baby Shampoos

Claims for the baby-care category refer to the safety of the products. Examples of popular claims in this area are "Tear-free," "Hypoallergenic," and phrases relating to the "gentleness" and "mildness" of the product, such as "Gentle cleansing for softer skin."

Eye-Sting Evaluations

For the claim "Tear-free," the clinical evaluation used for substantiation is the eye-sting test (76–78). The objective of this clinical study is to evaluate and compare the perceived discomfort and/or eye-sting potential of the test baby product.

Subject Selection.   Eye-sting evaluations are performed on adults, as they are better able to discriminate different levels of discomfort and/or eye sting.

The inclusion and exclusion criteria are essentially the same as those of the 90-day in-home use study subject selection requirements, with the addition of the following to the exclusion criteria:

Individuals who are wearers of contact lenses

Individuals who exhibit stinging (>2 rating) to a preevaluation with isotonic saline solution

Individuals with eye conditions as determined by ophthalmologist on staff

Preevaluation.   Prior to the start of a study, a preevaluation with isotonic saline is conducted. Each subject receives one drop of isotonic saline solution in the right eye first and rates the intensity of stinging/discomfort. After the subject finishes the rating of the right eye, the same procedure is followed for the left eye.

Stinging/discomfort Scale
0 = No stinging/discomfort
1 = Mild stinging/discomfort
2 = Moderate stinging/discomfort
3 = Severe stinging/discomfort
4 = Unbearable stinging/discomfort

A subject who experiences a stinging/discomfort intensity >2 with either the right or left eye is disqualified from the study.

The eyes of each subject are then evaluated and qualified by an ophthalmologist for visual acuity and the absence of conjunctival irritation, injection, and scleral redness.

Evaluation.   All qualified subjects are assigned to two test shampoos according to a randomization schedule. One drop (0.033 ml) of a 20% aqueous solution of the test shampoo is instilled in the right eye. The subject is then asked to rate the intensity of perceived stinging/discomfort within 10 s following instillation. The eye area is wiped with a water-saturated cotton ball and blotted dry. The subject will wait until the stinging/discomfort (if any) disappears prior to the left eye evaluation. The same procedure is followed for the left eye.

Following the completion of the evaluations, the eye area is examined again by the ophthalmologist.

Preclinical Safety Data.   All test products must have preclinical eye safety data prior to the initiation of this evaluation. The Primary

Draize Eye Irritation test is the accepted industry standard method for this purpose.

Results. The results of an eye-sting evaluation used in claims substantiation will indicate stinging/discomfort scores in the "none" to "moderate" range.

SENSORY METHODS

Introduction

While instrumental and clinical techniques provide important information in substantiating a particular product benefit, the success of a product in the marketplace requires the end users to recognize and believe in the benefit of the product. The assumption that since there is biological efficacy (e.g., silicone deposition found on hair), there will also be perceived efficacy by the consumer (e.g., product conditions the hair) is not true in many cases. It is important to understand these differences in efficacy to avoid negative consequences (79).

Sensory methods can be used in substantiating product performance claims. In this section, claim substantiation methods with trained (or expert) panels and naive (or consumer) panels are reviewed.

Trained or Expert Panel Testing

Trained or expert panels are used when the claim, either superiority or parity, involves a specific attribute (80). For example: "Hair spray X has more hold than hair spray Y" or "Removes tangles as well as brand X without buildup." Attributes are objectively measurable by the trained panelists. In the hair care area, licensed cosmetologists may be trained to become expert panelists to evaluate products on human heads.

Trained panels can be used to discriminate between products. Highly trained panels (descriptive test panel) can rate intensities of attributes for products. In general, discrimination testing requires a minimum 30–50 panelists and a descriptive panel usually consists of 6–15 highly trained panelists (80,82).

In claim substantiation, trained panel evaluation should focus only on claim-related attribute(s) rather than a comprehensive product description. Claims such as "Gives more shine than brand X" or

"Removes tangles as well as brand X" can be substantiated by trained panel test. Trained panel should not be used to measure subjective responses (e.g., preference or degree of liking).

Panel Training

Discrimination Panel.   The recruiting, screening, selection, training, and validation of a trained panel are discussed in detail in publications from American Society for Testing Materials (ASTM) (81–84,91). The discrimination panel training should include the proper procedure to handle the sample, the proper way to use the ballot, and a discussion on the importance of adhering to test protocol. Reference samples with various levels of difference to be evaluated, such as low to high "hold" for hair spray on tresses, should be provided to the panelists. This is especially important for the attribute difference tests.

Descriptive Panel.   The training of a descriptive panel is a much more involved process and takes a great deal of time and effort. The potential panelist has to successfully pass a battery of acuity tests to demonstrate that he or she cannot only detect the differences between products but also describe the differences in terms of character and in intensity. The training phase includes terminology development among the group for that product category, intensity ratings with anchors and references, pilot testing, and remedial training. The panel's performance should be continuously monitored and the progress communicated to the panel. A motivation program should be instituted to assure the best panel performance (82).

Consumer Testing

The success of a hair care product depends largely on how well the consumer responds to the claim. When the consumer perceives the product as having the claimed benefit, the likelihood of the market success is greatly increased.

Consumer testing is used to measure product users' subjective reaction, either liking or preference, to a product. This subjective reaction can be either attribute-specific or an overall opinion of the product. Both measurements can support superiority ("The best shampoo money can buy") or parity ("Leaves hair as shiny as brand X but costs less").

After the objective of the test is clearly identified, *only* those questions that relate to the claim should be asked in the questionnaire. Consumer should not be surveyed on responses beyond the scope of the claim (91–93).

Consumer testing data are usually key elements in the claim substantiation program for a hair care company. This subject is also discussed in detail in Chapter 9. In this section, the following consumer testing issues are briefly discussed.

Home Use Testing

Home use testing is usually the method of choice for hair care products. The advantage of the home use test is that product usage is under more typical condition. There is also opportunity for repeated experience. The drawbacks include the lack of control of usage, lack of assurance of the actual use of the product, and the reliance on respondents' ability to recall.

Central Location Testing

This method may be useful for products such as salon-oriented products. Well-equipped beauty salons in different demographic areas may be used and commercial stylists can be temporarily employed to conduct the test. Many companies also use central location test, such as mall intercept, to test consumer's perceptions on certain attributes, e.g., the fragrance of the product. Central location testing provides better control of product application and usage as well as immediate feedback.

Research Guidance Testing

Research guidance testing (RGT) is a consumer test with specialized functions. It is used to provide guidance to the product development process by asking the consumer both hedonic (liking) and intensity questions on product attributes. It can use either the "home use" or "central location" format among product users to obtain attribute specific information. This type of testing provides consumer-perceived attribute-intensity measurements that may be used in conjunction with other supporting evidence as part of the claim substantiation.

Test Design

To meet the objectives of substantiating a specific claim, a sound test design is critical. In addition to a clearly defined test objective and appropriate panelists (trained panelists for discrimination and descriptive tests and naive panelist and/or product users for consumer test), special attention should be paid to sample and substrate preparation, questionnaire design, data analysis, and results interpretation.

Sample Procurement

When products involved in the test are commercially available, they should be purchased from high-volume stores. The manufacturer's product should also go through the normal distribution channel prior to testing.

   If the product is not yet on the market, it should represent commercial production at the time of testing. It is desirable but may not always be practical for the new product to be made in the production facility. If a pilot plant batch must be used for claim support, then supplemental testing (e.g., a discrimination test for similarity) should be considered to demonstrate that the claimed benefits extend to material made at the production facility (80).

Sample Preparation

   Trained Panel.   All the samples for testing should be prepared in an uniform manner. Shampoo and conditioner should be measured and dispensed from a syringe or repeater pipette. The *ASTM Standard Practice for the Descriptive Analysis of Shampoo* (81) recommends 1.0 ml of shampoo for hair tresses weighing 3–4 g. For the human half-head test, using 5 ml per side for the first shampoo and 2.5–3 ml if there is a second shampoo is suggested. The methods of product application for panel testing differ depend on the product category, but they should be consistent and uniform within one test. For claim substantiation purposes, the standard sample preparation methods should be as close to the normal consumer practices as possible.

   Consumer Panel.   The principles remain the same for consumer panels as for trained panels on the consistency and uniformity of sample preparation. Special attention should be paid to the nature of blindness of the test by using the same kinds of undecorated packages labeled only with codes (i.e., three-digit random numbers) for all samples. For liquid products, the size of orifice of the bottle should be appropriate to the consistency of the product and the orifice should be identical for each sample. When transferring content of the product to the blinded packages is not possible, as in the case of aerosol products, the sample unit should be completely wrapped to conceal its identity.

Substrate Preparation

Hair Tresses. Trained panels can evaluate test samples on hair tresses or on the human head. Virgin hair that has never been chemically treated is available in different shades and lengths from vendors (Demeo Brothers, NY; International Hair Import Products Inc., White Plains, NY). Sample hair is usually weighed and mounted to a tab or other fastening device. After it is cut to a specific length, it should be stripped with either a standard shampoo or a detergent solution to remove any residue.

The use of hair tresses for trained panel testing provides minimum variability as compared to using hair on the human head. Tress testing allows multiple-product evaluation in a test with minimum cost. For claims specifically made for certain hair types (e.g., permed, tinted, bleached), treating virgin hair chemically following standard procedures assures the consistency of the results. Tresses treated with artificial sebum (88–90) also can be used in substantiating claims (e.g., cleanability). The test control offered by tress testing significantly increases the precision of the test and is specially useful for parity claims.

Human Head. The advantage of testing on human heads is that it approximates the real world. State law may require a licensed cosmetologist to apply hair care products to human heads. Typically, under those circumstances, licensed cosmetologists would receive additional training on product evaluation to be qualified as trained panelists.

Subject (model) selection should be based on the test objectives and/or target population of the product. Age, sex, hair type (tinted, permed, frosted, bleached), scalp condition (clear, flaky), hair condition (normal, dry, oily), hair diameter (fine, medium, heavy), hair texture (smooth, rough), hair length (short, medium, long), and hair density (low, medium, high) should be considered as criteria when a claim is intended to target a specific segment of people. An informed written consent should be obtained from subjects prior to their participation.

For trained panel evaluation, two types of tests are usually performed on human heads: on the whole head or the half head. Since the panelists (cosmetologists) are trained on measuring attributes and intensity, the difference between the two tests is mostly a matter of economics: the half-head procedure can test two products on one subject (model), whereas only one product can be evaluated on one subject using the whole-head approach. In the half-head protocol, either sequential monadic (one-half of the hair is washed and evaluated in steps independent of the other side) or paired comparison (compare and rate each attribute between sides) can be used and should be determined by the test objective.

For consumer tests in a salon environment or home use situation, subjects should be selected from a defined category of users or potential users. Company-maintained subject databases should not be used in consumer tests for claim substantiation because they are not random samples. Again, attention should be paid to balancing the hair characteristics and other demographics between each treatment.

Experimental Design

The exact statistical design of each test depends on the test objective. All samples should be tested blind and labeled only with three-digit random codes. The order of presentation for multiple samples in the hair care product category is usually sequential monatic. The order (or left-right sides) of the presentation should be balanced to avoid context/contrast effect and position bias. Replications are strongly recommended in trained panel testing to account for individual panelist measurement error.

Questionnaire Construction

The use of trained panel testing for claims support should focus only on product characteristics specific to the claim. Although both the trained panel and consumer panel are capable of describing the entire profile of the product or of reacting to any product-related responses, it is strongly recommended that they not be questioned on items beyond the scope of the claim.

Trained Panel.

1. *Paired Comparison Method*. Panelists choose one of the two products that has more or less of a certain attribute. For example: "Which side (left or right) of the hair has more shine?" A forced-choice method should be used.

2. *Attribute Rating*. Trained panelists make a mark on a scale to indicate the intensity of the attribute for each product. An unstructured or semistructured line scale is recommended to minimize the bias associated with the use of numbers. For example; "Please comb the hair tress X and rate it on the scale below."

```
0                                                    10
+ ———————————————————————————————————————————— +
no tangle                          a great deal of tangle
```

Now, use another comb, and comb the hair tress X and rate it on the scale below.

```
0                                                    10
+ ———————————————————————————————————————————— +
no tangle                          a great deal of tangle
```

3. *Descriptive Analysis*. Highly trained panels can provide detailed product performance descriptions, which are useful in substantiating attributes of composite nature. For example: "Rate the following lathering attributes of sample X."

```
Amount of foam      0                                    10
                    + ———————————————————————————————— +
                    none                               high
Thickness of foam   0                                    10
                    + ———————————————————————————————— +
                    thin                              thick
Denseness of foam   0                                    10
                    + ———————————————————————————————— +
                    airy                              dense
Bubble size         0                                    10
                    + ———————————————————————————————— +
                    small                             large
```

Wetness of foam   0   10

Consumer Panel. There are three kinds of consumer questions that can be asked in the consumer panel for claim substantiation: liking/hedonic/acceptance, preference, or diagnostic. If an overall impression of the product is the basis of the claim, then an overall liking or overall preference question should be asked. If claims are based on specific attributes, direct questions should be asked in the form of attribute liking, perceived attributes intensity, or attribute preference.

1. *Liking/Hedonic/Acceptance: Overall or Attributes.* The scale recommended by the ASTM (80) is nine-point hedonic scale with an anchor word for each point. For example: "How do you feel overall about this shampoo?"

Like extremely
Like very much
Like moderately
Like slightly
Neither like nor dislike
Dislike slightly
Dislike moderately
Dislike very much
Dislike extremely

Attribute liking questions can use the same format. Such as, "How do you feel about the lather of the shampoo?" followed by the nine-point scale.

Other types of liking scales (five points, seven points) can also be used along with different anchor words. A consistent format (number of points, position and wording of anchors) throughout the questionnaire is critical in questionnaire design.

2. *Attributes Diagnostic Questions.* This type of question collects information on the perceived intensity/level of the attribute (e.g., intensity/level of bounce or body). Attribute diagnostic questions can use an absolute intensity scale ("none" to "extreme") similar to the scales used in descriptive analysis.

3. *Preference*: *Overall or Attributes Preference*. Preference questions ask the consumer to choose one out of the two (or more) samples after experienced them based on their impression overall or on specific attributes (e.g., "Which hair spray do you prefer for giving your hair shine?")

A combination of questions mentioned above is frequently used in many consumer studies. When multiple questions are involved, the most important questions should be positioned at the beginning of a questionnaire to avoid influence from other questions. In general, it is recommended to put the overall questions before any attribute questions, since the overall question is usually the most important.

Open-ended comments are often collected in the consumer tests. However, it is not recommended to use the comments to substantiate claims.

Data Collection and Analysis

Data should be analyzed according to the statistical design. Caution should be exercised when making a parity claim. When no statistically significant differences can be found between two samples on a particular attribute, it does not imply parity. For example, no differences are found between the two hair sprays on "hold" does not imply that they are in parity on "hold." Gacula (85,86) has detailed discussions on this issue and the proper statistical analysis methods for the parity claim data.

Test Facility

Laboratory setting and the environmental conditions are critical for valid and reproducible test results. Detailed discussions on the sensory laboratory facility design can be found in the publication *Physical Requirement Guidelines for Sensory Evaluation Laboratories* (94).

Special planning considerations in hair care product testing laboratories should include the following:

1. *Light*. An even and consistent light source is recommended. Simulated northern daylight should be considered, especially when hair color and hair shine evaluations are involved.

2. *Water*. Controlled water temperature, pressure, flow rate, and

hardness should be considered and continuously monitored to ensure the reproducibility of the results.

3. *Air.* Air temperature of 22°C (72°F) and 45–50% humidity are recommended in the ASTM manual. A quick and effective air exhaust system to remove the product's odor and fragrance is vital in hair care product testing. A slight positive air pressure in the testing room is also recommended to prevent inflow of air and odor from other areas.

4. *Electric Outlets.* Adequate electric outlets should be installed in the testing booths/stations to accommodate equipment such as dryers and/or computers.

Consumer Panel

Consumer panels are usually conducted in commercial salons or in subjects' homes. The important test control for the physical environment is that, for each subject, all the samples are applied and used under the same conditions (i.e., similar light, water temperature, and pressure) and handled the same way as instructed.

The product developer has a wide variety of methods available to support the truthfulness of these claims. Instrumental, clinical, and sensory techniques described in this chapter have all been successfully used in predicting consumer perception of product performance—a requirement for any claims testing program.

This chapter has given an overview of methods for each of the three classes of testing, all of which have been used in developing great hair care products and substantiating claims about them. Instrumental techniques that were described measure physical and combing properties, hair surface characteristics, deposition and penetration of conditioning agents, changes in hair's flexibility and rigidity, and hair shine. Clinical methods presented support claims commonly associated with shampoos and conditioners. These methods are used to substantiate antidandruff efficacy; to support claims that a product has been clinically, pediatrician-, or dermatologist-tested; and to demonstrate tear-free properties of shampoos. Sensory tests discussed show how trained and consumer panels can be used to assist in formulation development and to support claims about specific product attributes such as better foaming, more conditioning, easy rinsing, added luster, and no static.

Product developers are faced with the challenge of documenting product performance so that the aggressive claims being made for their hair care brands truly reflect product attributes. The testing methodology presented in this chapter can be used in developing a successful strategy for supporting claims for most products in the hair care category.

REFERENCES

1. Busch P. Subjective and objective methods in hair assessment. Arzt Kosmetol 1989; 19:270–315.
2. Busche MG. (1967) Materials Engineering, Special Report—Part 1.
3. Szadurski J, Erlemann G. The hair loop test—A new method of evaluating perm lotions, 12th IFSCC Congress, Paris, France, September 13–17, 1982.
4. Szadurski J, Erlemann G. The hair loop test—A new method of Evaluating Perm Lotions, Cosmet Toiletr 1984; 99:41–46.
5. Wickett RR. Kinetic studies of hair reduction using a single fiber technique. J Soc Cosmet Chem 1983; 34:301–316.
6. Evans T, Ventura T, Wayne A. The kinetics of hair reduction J Soc Cosmet Chem 1994; 45:279–298.
7. Schwartz A, Knowles DJ. J Soc Cosmet Chem 1963; 14:455–463.
8. Scott GV, Robbins CR. Effects of surfactant solutions on hair fiber friction. J Soc Cosmet Chem 1980; 31:179–200.
9. Fair N, Gupta BS. J Soc Cosmet Chem 1982; 33:229–242.
10. Horiuchi T, Kashiwa Y, Ohno K. J Soc Cosmet Chem Jpn 1980; 14:62–65.
11. Ohno K. Fragr J Jpn 1982; 10(3):85–87.
12. Robbins CR. Chemical and Physical Behavior of Human Hair, 2d ed. New York: Springer-Verlag, 1988:276–278.
13. Garcia LM, Diaz J, Combability measurements on human hair. J Soc Cosmet Chem 1976; 27:379–398.
14. Scott GV. Spectrophotometric determination of cationic surfactants with orange II. Anal Chem 1969; 40:768–773.
15. Scott GV, Robbins CR, Barnhurst JD. Sorption of quaternary ammonium surfactants by human hair. J Soc Cosmet Chem 1969; 20:135–152.
16. Crawford RJ, Robbins CR. A replacement for Rubine dye for detecting cationics on keratin. J Soc Cosmet Chem 1980; 31:273–278.
17. Robbins CR, Reich C, Clarke J. Dye staining and the removal of cation-

ics from keratin: The structure and the influence of the washing anion. J Soc Cosmet Chem 1989; 40:205–214.

18. Sandoz Testing Procedure: Detection of Cationic Deposition Using the Pyrazol Test. Florham Park, NJ: Sandoz Corp.
19. Carson JC. Hair conditioning from shampoos—A comparative study of the effects of quaternized hydroxyethyl cellulose polymers. Soaps Cosmet Chem Spec 1989; October: 30–31, 70–71.
20. Thompson WE, Mills CM. Proceedings on the Scientific Section, TGA. 1951; 15:12–15.
21. Den Beste M, Moyer A. J Soc Cosmet Chem 1968; 19:595–609.
22. Stamm RF, Garcia ML, Fucha JJ. J Soc Cosmet Chem 1977; 28: 571–609.
23. Robbins CR. Chemical and Physical Behavior of Human Hair, 2d ed. New York: Springer-Verlag, 1988:272–275.
24. Stavrakas EJ, Platt MM, Hamburger WJ. Determination of Curl Strength of Tresses Treated with Water, Hair Spray, and Waving Lotion. Toilet Goods Association, Proceedings of Scientific Section, 1959; 31:36–39.
25. Johnson SC, Murphy EJ. An Improved Method for the In-Vitro Evaluation of Hair Holding Products. Technical Bulletin. Wayne NJ: GAF Laboratories.
26. Micchelli AL, Koehler FT. Polymer properties influencing curl retention at high humidity. J Soc Cosmet Chem 1988; 19:863–880.
27. Takada S. Aerosol Rep 1972; 11:12–25.
28. Ayer RP, Thompson JA. J Soc Cosmet Chem 1972; 23:617–636.
29. GAF Testing Procedure # 101, Humidity Curl Retention Test. Wayne, NJ: GAF Laboratories.
30. Brookins MG. J Soc Cosmet Chem 1965; 16:309–315.
31. Ellison MS. Robot-based electrostatic field measurement system. J Text Inst 1991, 4:512–513.
32. ASTM Standard Test Method for Electrostatic Propensity of Textiles. Designation: D 4238-83:822–825.
33. ASTM Standard Test Method for Static Electrification. Designation: D 4470-86:435–440.
34. ASTM Tentative Recommended Practice for Electrostatic Charge Mobility on Flexible Barrier Materials. Designation: F 365-73:360–365.
35. Mills CM, Ester VC, Henkin H. Measurement of Static Charge on Hair. J Soc Cosmet Chem 1956; 7:466–475.
36. Lunn AC, Robert EE. The electrostatic properties of human hair. J Soc Cosmet Chem 1977; 28:549–569.
37. Wis-Surel G, Jachowicz J, Garcia M. Triboelectric charge distributions

generated during combing of hair tresses. J Soc Cosmet Chem 1987; 38:341–350.

38. Puri AK, Jones RT. An Approach to Permanent Hair Conditioning. XIV IFSCC Congress, Barcelona, Spain. 1986; 2:1153–1155.

39. Jones RT. Substantivity and Properties of Wheat Protein as Applied to Hair. Technical Communication. Edison, NJ: Croda, Inc.

40. Jones RT. A Study of the Substantivity and Penetration of Crotein HKP into Hair. Private correspondence. Parsippany, NJ: Croda Inc, 1993.

41. Hair Deposition, Penetration and Moisturization. Private correspondence. Stamford, CT: Dermatech of Connecticut Inc, 1993.

42. Ayer RP, Thompson JA. Scanning electron microscopy and other new approaches to hair spray evaluation. J Soc Cosmet Chem 1977; 23:617–636.

43. Brown AC, Swift JA. New Techniques for the Scanning Electron Microscope Examination of Keratin Fibre Surfaces. Proc. Fifth European Congress on Electron Microscopy, 1972:386–387.

44. DiBlanca S. Innovative scanning electron microscopic techniques for evaluating hair care products. J Soc Cosmet Chem 1973; 24:609–622.

45. Puderbach H, Flemming P. X-ray analysis on the scanning electron microscope in hair cosmetic evaluation and development. Cosm Toiletr, 1979; 94:79–84.

46. Riso R. Surfactant effects on the hair, Part 1. D&CI 1973; October:39–43, 133–134.

47. Riso R. Surfactant effects on the hair, Part 2. D&CI, 1973; November:46–50, 112B.

48. Robbins CR, Crawford RJ. Cuticle damage and the tensile properties of human hair. J Soc Cosmet Chem 1991; 42:59–67.

49. Swift JA. Fine details on the surface of human hair. Int J Cosmet Sci 1991; 13:143–159.

50. Swift JA, Brown AC. The critical determination of fine changes in the surface architecture of human hair due to cosmetic treatment. J Soc Cosmet Chem 1972; 23:695–702.

51. Tate ML et al. Quantification and prevention of hair damage. J Soc Cosmet Chem 1993; 44:347–371.

52. Goldstein JI, et al. Scanning Electron Microscopy and X-ray Microanalysis, 2d ed. New York: Plenum Press, 1992.

53. Gabriel BL. SEM: A User's Manual for Materials Science. Metals Park, OH: ASM, 1985.

54. Cooperman ES, Johnsen VL. Penetration of protein hydrolysates into human hair strands. Cosmet Perfum 1973; 88:19–22.

55. Wickett RR. Kinetic studies of hair reduction using a single fiber technique. J Soc Cosmet Chem 1983; 34:301–316.

56. Evans DJ. A method for determining the penetration of reducing agents onto wool using fluorescence microscopy. Textile Res J 1989; 59:569–576.

57. Evans T, Ventura T, Wayne A. The kinetics of hair reduction. J Soc Cosmet Chem 1994; 45:279–290.

58. Weigmann HD, et al. Characterization of surface deposits on human hair fibers. J Soc Cosmet Chem 1990; 41:379–390.

59. Jurdana LE, Leaver IH. Characterisation of the surface of wool and hair using microscopical and fluorescence probe techniques. Polymer Int 1992; 27:197–206.

60. Haughland R. Molecular Probes, Handbook of Fluorescent Probes and Research Chemicals, 5th ed. Eugene, OR: Molecular Probes, Inc, 1992.

61. Vacca L. Laboratory Manual of Histochemistry. New York: Raven Press, 1985.

62. 21 Code of Federal Regulation, Food and Drug Administration, Parts 310 and 358, Dandruff, Seborrheic Dermatitis, and Psoriasis Drug Products for Over-the-Counter Human Use; Final Rule.

63. Botwinick CG, Botwinick I. Methods for Evaluating Antidandruff Agents, Proceedings of the Scientific Section by the Toilet Goods Association, 47:17, 1967.

64. Orentreich N, Taylor EH, Berger RA, Aberbach R. Comparative Study of Two Antidandruff Preparations. J Pharm Sci 1969; 58:1279.

65. Van Abbe NJ, Dean PM. The clinical evaluation of anti-dandruff shampoos. J Soc Cosmet Chem 1967; 18:439.

66. Van Abbe NJ. The investigation of dandruff. J Soc Cosmet Chem 1964; 15:690.

67. Finkelstein P, Laden K. An objective method for evaluation of dandruff severity. J Soc Cosmet Chem 1968; 19:669.

68. Troller JA. Model system for the investigation of dandruff. J Soc Cosmet Chem 1971; 22:187.

69. Laden K. Comparative chemical study of dandruff flakes, skin scrapings and callus. J Soc Cosmet Chem 1965; 16:491.

70. Plewig G, Kligman AM. The effect of selenium sulfide on epidermal turnover of normal and dandruff scalps. J Soc Cosmet Chem 1969; 20:765.

71. McGinley KJ, Marples RR, Plewig GA. A method for visualizing and quantitating the desquamating portion of the human stratum corneum. J Invest Dermatol 1969; 53:107.

72. Essex Testing Clinic. Clinical Efficacy Evaluation of an Anti-Dandruff Shampoo Protocol. Verona, NJ: Essex Testing Clinic

73. Meltzer N. Society of Cosmetic Chemists Presentation, Kathon 90-Day Patch Use Patch, December 1980.

74. Code of Federal Regulation, 47: No. 233. Dec 3, 1982, p 54658.

75. Research Testing Laboratories. A Twelve Week Normal In-Home Usage Study to Determine the Efficacy and Safety of Shampoo Products Protocol. Great Neck, NY: Research Testing Laboratories.

76. Shanahan RW, Ward CO. An animal model for estimating the relative sting potential of shampoos. J Soc Cosmet Chem 1975; 26:581.

77. Van Abbe NJ. Eye irritation: Studies relating to responses in man and laboratory animals. J Soc Cosmet Chem 1973; 24:685.

78. Essex Testing Clinic. Clinical Evaluation of the Relative Eye Sting Potential Protocol, Verona, NJ: Essex Testing Clinic.

79. Stone H, Sidel JL. Perceived Efficacy and Advertising Claims: Sensory Evaluation Practices, 2d ed. San Diego, CA: Academic Press, 1993: 279–282.

80. Advertising Claim Substantiation, ASTM Task Group E18.08.07 (draft 1994).

81. Standard Practice for the Descriptive Analysis of Shampoo, ASTM Task Group E18.03 (draft 1994).

82. Meilgaard M, Civille GV, Carr BT. Sensory Evaluation Techniques, 2d ed. Boca Raton, FL: CRC Press, 1991.

83. Guidelines for the Selection and Training of Sensory Panel Members, ASTM Special Technical Publication (STP), 758, 1981.

84. Hootman RC. Manual on Descriptive Analysis Testing for Sensory Evaluation, Philadelphia: ASTM MNL 13, 1992.

85. Gacula MC Jr. Claim Substantiation, Design and Analysis of Sensory Optimization. Trumbull, Connecticut: Food and Nutrition Press, 1993: 237–255.

86. Gacula MC Jr. Claim substantiation for sensory equivalence and superiority. In: Lawless HT, Klein BP eds. Sensory Science Theory and Applications in Foods. New York: Marcel Dekker, 1991:413–436.

87. Munoz AM, Civille GV, Carr BT. Sensory Evaluation in Quality Control. New York: Van Nostrand Reinhold, 1992.

88. Clarke J, Robbins CR, Schroff B. Selective removal of sebum components from hair by surfactants. J Soc Cosmet Chem 1989; 40:309–320.

89. Clarke J, Robbins CR, Schroff B. Selective removal of sebum components from hair: II. Effect of temperature. J Soc Cosmet Chem 1990; 41:335–345.

90. Thompson D, Kenaster C, Allen R, Whittam J. Evaluation of relative shampoo detergency. J Soc Cosmet Chem 1985; 36:271–286.
91. Read J. Role of Marketing Research in Claim Testing. Food Tech 1994; August:75–78.
92. Edelstein JS. Supporting and challenging advertising claims with consumer perception studies. Food Tech 1994; August:79–82.
93. Passman N. Supporting advertising superiority claims with taste test. Food Tech 1994; August:71–74.
94. Egret J, Zoo K. Physical Requirement Guidelines for Sensory Evaluation Laboratories. Philadelphia: ASTM MNL, 1986: 913.

3

Cutaneous Biometrics and Claims Support

Mark Willoughby and Howard I. Maibach
University of California, San Francisco, California

INTRODUCTION

The development and acceptance of bioengineering techniques to quantitate skin function has revolutionized the evaluation of personal care products. Many investigations were plagued by inconsistencies related to visual assessments. The previous generation of studies were based on an observer's visual scaling and relied on remembering the previous status of the skin, so that claims of improvement or exacerbation could be made. Because evaluation scales varied between investigators, it was difficult to relate different studies to each other. Objective methods now quantify skin parameters, including transepidermal water loss (TEWL), skin surface water loss (SSWL), skin hydration, pH, color, capillary blood flow, thickness, casual sebum level, and so on and document the mechanical, physical, and functional properties of the skin. These methods are noninvasive, reproducible, and sufficiently sensitive to detect changes not apparent visually.

Recent implementation by investigators of several statistical models have made the evaluation of personal care products facile and reproducible. Large sample size can be avoided by using volunteers as their own control in crossover design and bilateral paired comparison assays as well as using randomization, blocking, and factorial designs.

The combination of improved experimental design and the ability to quantitate the evaluation of personal care products enables investigators to perform small but discriminating evaluations. The value of bioengineering is not as a substitute for traditional clinical trials but as an addition to them. In the end, the consumer must be the validation instrument.

EXPERIMENTAL DESIGN

Goals

This section focuses on efficient experimental design. The ideal design has three major characteristics: it minimizes bias, it is maximally efficient with resources and information, and it decreases confounding (1). The term *confounding* describes the situation in which it may be impossible to distinguish treatment effects from extraneous factors (2,3).

Clinical Trial Designs

The design of clinical trials has recently evolved. Current studies are often constructed to be double-blind, placebo-controlled, and randomized. Studies not meeting these requirements are often quickly rejected.

A major obstacle, often preventing the achievement statistically and clinically significant results, was the impracticality of large sample size. Without a large sample size, the possibility that confounding would significantly alter the results could not be ruled out. More recently, however, studies have been designed in which the volunteer acts as his or her own control. Having subjects act as their own controls not only effectively doubles the number available but also minimizes the intrusion of such potentially confounding factors as age, gender, and environment (1).

Different experimental designs effectively utilize volunteers as their own controls to vastly improve both the ease of performing the experiment and the significance of the results. These specialized experimental designs include crossover studies and bilateral paired comparisons. Crossover studies, in which the volunteer receives the opposite treatment once the effect of the first treatment has cleared, are limited to evaluations of agents intended for symptomatic relief of conditions like xerosis, which revert to their pretreatment state shortly after discontinuing therapy (2). Bilateral paired comparisons require that the patient have approximately symmetrical lesions of equal severity in nearly identical locations on each side of the body (as in psoriasis) and that the medication cannot significantly migrate or be accidentally transferred from one side to the other (2,4).

Because of the strict conditions (e.g., having a condition that is not permanently altered by the treatment or altered over the period of time required to complete the study) required in designing a crossover study or a bilateral paired comparison, these designs are not always an option for the clinician treating curable disease. Instead, the investigator must thoughtfully design an experiment utilizing statistical techniques aimed at reducing the effect of confounding. These techniques include randomization, simple and paired comparisons, blocking, and the utilization of factorial designs.

Randomization

Randomization is aimed at eliminating subconscious bias on the part of the investigator as the subjects are separated into groups. A prevalent belief is that randomization makes the groups approximately equal in all respects except the treatments being compared; however, this is seldom the case unless the number of patients involved is large (1). Simple randomization creates the potential for confounding, since, by chance, a higher proportion of patients with more severe disease or some other adverse prognostic factor may be allocated to one group rather than another (2,3).

Simple Comparisons

Simple comparisons are the most basic of experimental designs. In a simple comparison, the subjects are divided into two equally sized groups permitting the comparison of the differential effects of the treatments. Each group should have approximately the same number of subjects and the process of dividing the subjects into two groups should be randomized. Using equally sized groups ensures that the statistical analyses will produce the most efficient results possible considering the number of subjects selected. After dividing the subjects into groups A and B, treatment 1 is randomly assigned to one group and treatment 2 to the other. A "treatment" does not necessarily need to be a medical treatment. For example, a placebo, a vehicle, or even no treatment at all is spoken of as allocation of a "treatment" (2). Random allocation of subjects into groups (e.g., by telephone numbers or home addresses) must be performed to ensure that the clinician does not insert his or her own bias into choosing group members. Subconscious bias may, for example, lead the investigator to allocate the patients with less severe cases to the new treatment, making it impossible to determine whether a difference in outcomes is due to a difference in efficacy (2).

Paired Comparison

Paired comparisons are established by pairing subjects with similar attributes that may later be a source of confounding. Then each subject of the pair is randomly assigned to one of the treatment groups. The special advantage of pairing is that differences between treatment groups can be more safely ascribed to the effects of treatment than to interfering effects from differences in such factors as disease type or severity (2). Paired comparisons also tend to be more efficient statistically than unpaired comparisons (5). The bilateral paired comparison described above is an excellent example of how pairing can limit the possibility of confounding.

Randomized Blocks

Blocking utilizes the same principles as pairing to reduce the chance of confounding. Subjects sharing the same attributes that may later be a source of confounding are grouped together and referred to as a block. Grouping into blocks is done before the sample is randomized. The block design automatically balances the numbers of subjects of each type in each treatment group, thereby simplifying analysis and interpretation (1). In effect, utilizing a randomized block design allows the clinician to treat each block as a small trial (Fig. 1).

The randomized block design permits comparison of any number of treatments in the presence of one extraneous source of variability. The randomized block design produces more and better information than a completely randomized design for several reasons (2). First,

		Clinic A								Clinic B		
		TR	TM	EF						TM	EF	
Patient	1	H	H	R			1	R	H	R		
Number	2	H	R	H			2	R	R	H		
	3	R	R	R			3	H	H	H		
	4	H	R	H			4	R	H	H		
	5	R	H	H			5	H	R	R		
	6	R	H	R			6	H	R	R		
Block No.		I	II	III				I	II	III		

Figure 1 Randomized block design in a trial of two topical antifungals. (From Ref. 1.)

treatment comparisons are made between like subjects within a single block, ensuring that the results are not due to confounding factors, such as age or gender. Second, design efficiency is optimized, as each treatment is administered equally as often within each block. Third, it is possible to establish whether treatment outcomes are due to age or gender, since each age-gender group (block) has been treated in the same way. These advantages make the randomized block design one of the most powerful tools in designing an efficient trial.

Factorial Designs

Factorial experiments are useful in examining the effects of multiple factors on a single response. The factorial design allows one to obtain both the single effect of a treatment and the treatment's interaction with other treatments.

The simplest example of factorial design, the 2 × 2 design, is displayed in Fig. 2 (1). Half of the patients get drug A and half get drug B. The division into treatment groups is such that all possible combinations exist in equal numbers, accounting for the 2 × 2 architecture. The main effect of each drug is determined by subtraction. Interaction is present if the combined effects of treatment are significantly greater or lesser than the independent contributions of the separate effects (1).

Thus, factorial design serves not only to determine the primary effects of a particular treatment with the fewest number of subjects but

AS	A
A	O

Effect of   AS-A = Effect of corticosteroid in the presence of antibiotic
Corticosteroid   S-O = Effect of corticosteroid in the absence of antibiotic

Effect of   AS-S = Effect of antibiotic in the presence of corticosteroid
Antibiotic   A-O = Effect of antibiotic in the absence of corticosteroid

Figure 2   Two-by-two (2×2) factorial design in evaluating two treatments for secondarily infected chronic dermatoses. (From Ref. 1.)

also to measure the interaction (synergism or antagonism) between two or more treatments (1).

Increasing Efficiency of Design

Sample Size

The efficiency of an experiment depends upon selecting an appropriate sample size. In reality, a trial might consist of over 1000 subjects so as to dismiss the possibility that sampling error has affected the results. However, large trials are rare and may often be impossible due to the need for subjects with a specific condition. Even if it were theoretically possible to perform a certain experiment with 1000 patients, impracticality prevents them from being performed.

The vast majority of clinical trials are relatively small for practical reasons. Problems result from performing small trials. They come in two forms: first, there is the likelihood that competing treatments will not be recognized as being substantially different in efficacy if in fact they are; second, subgroups of patients who respond differently from the usual patient with the disease will not be recognized as being different in this regard (6).

Another factor resulting in small trial sizes is the erroneous belief in the law of small numbers (7). Briefly, this law states that those in the scientific community have an inherent belief that small samples are representative of the populations from which they are drawn. This faith is totally unjustified; small samples can be unrepresentative and, not surprisingly, often are (7).

Measurement Scales

Ideal scales of measurement for clinical trials are linear (divided into equal intervals and able to be added, subtracted, multiplied, and divided), standardized (universally understood), clinically relevant, and readily understood (1,8). However, this type of scale applies only to quantitative measurements.

The more common scale in clinical trials is an ordinal scale. Ordinal scales are semiquantitative scales: conditions are classified by order of severity (none, mild, moderate, severe) and therapeutic responses are categorized by degree of improvement (none, little, moderate, marked) (1). Ordinal scales cause problems in analyzing data and making comparisons between studies. As ordinal scales are nonlinear, it is inappropriate to add and subtract results. For example, it is erroneous to conclude that moderate xerosis is twice as severe as mild xerosis. Problems also arise due to the fact that it is difficult to standardize ordinal scales. Where one clinician may see a condition as markedly improved, another clinician may find only moderate improvement.

Fortunately for clinicians in the field of personal care products, new instruments are constantly being introduced that yield linear, quantitative results. Quantitative results allow the clinician to use standardized, linear scales that are readily understood.

INSTRUMENTAL EVALUATION

Transepidermal Water Loss

Water diffusion through the stratum corneum is measurable at the skin surface as transepidermal water loss (TEWL) in micrograms per square centimeter per hour. In vivo, the concentration gradient depends upon relative humidity and the skin temperature (9). As skin temperature increases, TEWL increases. Hydration of the stratum corneum, increasing the membrane diffusion constant, increases TEWL (10). Transepidermal water loss reflects the integrity of the water barrier function of the stratum corneum and is utilized, for example, to assess skin irritation (11).

In vivo, TEWL can be measured utilizing several water sampling techniques (12), as outlined below.

1. *Closed Chamber Method.* This consists of a capsule applied to the skin, collecting the vapor loss from the skin surface. The relative humidity inside the capsule is recorded with an electronic hydrosensor. The change in vapor loss concentration is initially fast and decreases proportionally as the humidity approaches 100%. The closed chamber method does not permit recording of continuous TEWL because when the air inside the chamber is saturated, skin evaporation ceases.

2. *Ventilated Chamber Method.* A chamber through which a gas of known water content is passed is applied to the skin. The water is picked up by the gas and measured through a hydrometer. This method allows the measurement of transepidermal water loss continuously; but if the carrier gas is too dry, it artificially increases evaporation.

3. *Open Chamber Method.* The open chamber method utilizes a skin capsule open to the atmosphere. The TEWL is calculated from the slope provided by two hydrosensors precisely oriented in the chamber. Air movement and humidity are the greatest drawbacks of this method when in vivo studies are performed.

Two instruments are currently accepted and used for the measurement of TEWL and both are commercially available: the Evaporimeter EP1 (ServoMed, Stockholm, Sweden) and the Tewameter (Courage & Khazaka, Cologne, Germany; Acaderm, Menlo Park, CA). The probe of the Evaporimeter includes an open chamber 12 mm in diameter mounted with sensors for determination of temperature and relative humidity. The TEWL is then automatically computed and displayed in grams per square meter per hour. The Tewameter uses the vapor pressure gradient to compute the TEWL.

The TEWL measurements correlate with impairment of the barrier function, but besides individual and racial (13) variability, the elicited responses vary from irritant to irritant and do not allow comparison between different molecules unless closely related. Moreover, TEWL may detect and monitor invisible irritation and aid in discriminating between between irritant and allergic reactions (14,15). In allergic reactions, the water barrier is initially not damaged if the application time of the test substance is short (that is, 24 h); the TEWL increases after 48 h, when the damage is secondary to the inflammatory infiltrate. Irritant reactions, on the contrary, show earlier increase of TEWL values due to barrier damage.

Variation in TEWL measurements were thoroughly reviewed by Pinnagoda et al (16). The TEWL varies according to number of variables unique to the individual, the environment, and the proper use of the instruments. Baseline TEWL is, for the major part of the range, age-independent. Premature infants have increased TEWL during their first weeks, and elderly skin may show decreased TEWL. Anatomic site is an important variable with respect to baseline TEWL, which can be ranked as follows: palm > sole > forehead = postauricular skin = nail = dorsum of hand > forearm = upper arm = thigh = chest = abdomen = back. However, neither gender nor race appears to have an effect on baseline TEWL. The intraindividual variation of baseline TEWL is considerably less than the interindividual variation by sites and by days. Physical, thermal, or emotional sweating are important variables to control. The TEWL does not appear to be influenced by simple vasonconstriction and vasodilation. Skin surface temperature is important for TEWL, and preconditioning of the test subject is required. Skin surface temperature should be measured and reported in publications, particularly if ambient room air temperature deviates from 20–22°C. The TEWL values also vary according to other environmentally related variables: air convection, ambient air temperature, geographic variation due to differences in ambient air humidity and temperature, and differences in water vapor partial pressure at different altitudes and geographic locations. Instrument-related variables must be controlled to achieve consistent and reproducible TEWL measurements. Abrupt changes in humidity or temperature of the probe or direct light on the probe may prevent the Evaporimeter from rezeroing between measurements. Maintenance of a horizontal measuring plane and constant, light pressure of the probe against the skin are required for consistent results.

Skin Surface Water Loss

Skin surface water loss (SSWL) represents the water evaporation from the skin surface in the special case of occlusion or after application of

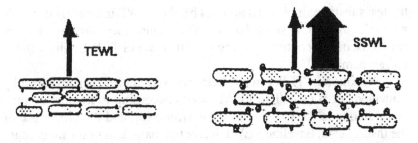

Figure 3 Schematic representation of baseline transepidermal water loss (TEWL) on the left side and the skin surface water loss (SSWL) of an artificially hydrated stratum corneum. In the case of (hyper-) hydrated stratum corneum as a result of occlusion or application of an aqueous solution, SSWL equals baseline TEWL plus excess water evaporation. (From Ref. 17.)

water or an aqueous solution to the skin. In essence, SSWL equals excess water evaporation plus baseline transepidermal water loss (Fig. 3) (17). The SSWL decay curves relate to the capacity of the stratum corneum to hold water and are lower after removal of lipids because a lower amount of bound water is available for evaporation (18).

Stratum Corneum Water Content

Recently, interest in the water content of the stratum corneum has increased considerably, since it influences various physical characteristics of the skin—like viscoelastic properties and functional characteristics such as barrier function and drug penetration. The diffusion of the water from the body to the stratum corneum is an equilibrium between the environment and the deeper skin layers. There is a concentration gradient within the stratum corneum which results in a continuous diffusion of water from the body to the skin and to the environment (TEWL) (19).

Methods have been developed to study the water content of the stratum corneum:

1. Measurements via impedance, resistance, and phase shift can be made with several electrical devices using different frequencies and

technologies. Electrical conductance, indicating the hydration state of the superficial epidermis, can be measured by a Skicon-100 (Acaderm, Menlo Park, CA) high-frequency hydrometer and displayed digitally as reciprocal impedance ($1/\mu\Omega$) (20–22). Capacitance as a measure of stratum corneum hydration can also be measured using a capacitance meter (Corneometer CM 820 PC; Courage & Khazaka, Cologne, Germany). The Nova Dermal Phase measures capacitance by determining the phase shift in alternating current and is particularly useful for measuring skin moisture levels after moisturizer treatments (23). Figure 4 (adapted from J.E. Wild) shows the results of an effective moisturizer treatment in which increased moisture content is shown by higher meter readings. The test, described as a desorption study, can be used to measure the efficacy of a moisturizer treatment over time (24).

2. The Microwave Probe (Acaderm, Menlo Park, CA) can be used to evaluate water content while avoiding the influences on the measures by ions or charges on the cell membranes (25).

3. Photoacoustic methods of measurement are based on the recording of the acoustic signal arising in the tissue following pressure variations caused by periodic light radiation (26).

4. Fourier transform infrared (FTIR) and attenuated total reflectance (ATM) techniques measure levels of lipids important to skin barrier function (23).

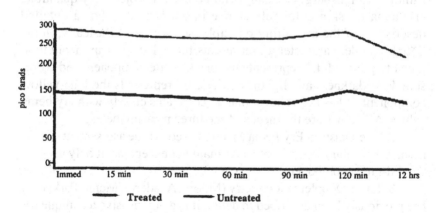

Figure 4 Measurements taken at timed increments following a single treatment with test sample versus nontreatment site. (From Ref. 23.)

Skin Friction

The Newcastle and the more recent friction meters (Acaderm, Menlo Park, CA) have been used to measure skin friction, one of the primary traits perceived by the consumer (23). It contributes to many of the other consumer-perceived descriptive attributes of the skin, such as roughness, scaliness, and flaking (27). Skin friction instrumentation is most often used to assess the effects of moisturization.

Skin pH

Skin pH can be measured with a flat glass electrode and a pH meter (Skin pH meter pH 900, Courage & Khazaka, Cologne, Germany). The pH value represents the presence of water-soluble components on the skin surface and may be indicative of dehydration effects or other changes taking place in the superficial skin (23,28).

Skin Color/Capillary Blood Flow

In vivo, skin redness (capillary blood flow) or overall skin color can be measured using three different techniques:

1. The Minolta chromameter, a tristimulus chromameter (CR 200, Minolta, Ahrensburg, Germany) can be used to objectively quantitate erythema by skin color reflectance measurements. (For a detailed description of the measuring principle, see Ref. 29.) Basically, in the L*a*b* mode, parameter a* represents the color spectrum from total green to pure red. L* represents the black-white component and corresponds to darkness and lightness, while b* represents the blue-yellow component. It has been shown that a* correlates closely with erythema values. All values are the means of the three measurements.

2. The Diastron Erythema Meter is used for the assessment of redness. It is similar to the Minolta chromameter except that it only uses a*, the red-green component.

3. Laser Doppler velocimetry (Moore, Acaderm, Menlo Park, CA) has previously been described in detail; it is a noninvasive technique for measuring blood-flow changes in the microcirculation of the skin (30,31). A monochromatic light from a helium-neon laser at 632.8 nm

is transmitted through optical fibers to the skin. The light is reflected with Doppler-shifted frequencies from the moving blood cells in the upper dermis at the depth of approximately 1 mm. The laser Doppler velocimeter extracts the frequency-shifted signal and derives an output proportional to the flux of erythrocytes in the blood flow (30). The shift increases with increasing velocity. In a mechanical model simulating the microvascular pattern of the skin, a linear relationship between laser Doppler velocimetry and blood flow was detected for low and moderate flow rates (32). For higher flow, photo multiple scattering and increased light absorption due to higher erythrocyte volume fraction cause a slight underestimation of flow. Laser Doppler velocimetry seems useful in discriminating between negative and positive reactions but fails to quantify strongly positive patch-test reactions (33). Indeed, in a series of 31 patch tests evaluated with both laser Doppler velocimetry and visual scoring, Staberg and associates found a good correlation between the two methods, reporting a fivefold increase in blood flow in sites scored "doubtful" and a tenfold increase in those scored "positive" (33,34). Laser Doppler velocimetry may be useful in evaluating the degree of skin irritation (13,32). Although laser Doppler velocimetry can be used to quantify the strength of allergic and irritant skin reactions, the technique cannot discriminate between the two (34).

Skin Surface Topography

Profilometry, image analyses, and ultrasound represent highly specialized techniques for the evaluation of skin conditions; instruments such as the Scopeman or Microwatcher (Acaderm, Menlo Park, CA) are designed to enhance and assist in the optical collection of skin surface images (23). These instruments are used to evaluate the effects of moisturizers, surfactants, acne medications, and the like (35–37).

Skin Reflectance Spectrophotometry

The modified reflectance spectrophotometer has been described in detail elsewhere (38). Briefly, polychromatic light from a xenon short-arc lamp (Osram XBO 1500) is guided to the skin through a flexible light guide (Hirschmann, Germany). Skin reflectance is collected in an integrating sphere and guided through a similar fiber to a monochromator (Jobin Yvon H20, France), which splits the light into 5-nm bands in the spectral range 355–700 nm. Skin reflectance is detected by a photomultiplier (Hammamatsu, Japan) and analog-to-digital converted for further processing and computer (IBM-AT) storage. Each chromaphore is detected by analyzing skin reflectance within specific ranges of the measured spectrum (355–700 nm) (39). Initially, melanin content was calculated in the range 360–390 nm, which corresponds to the main in vitro absorbing area in the recorded spectrum. Oxygenated and deoxygenated hemoglobins were subsequently analyzed using all data points from 515 to 610 nm and incorporating the previously determined melanin content. By linear regression analysis, the optical amounts (c_x × d_x) of the chromophores were calculated, where c represents the concentration and d the optical distance (thickness) of the chromophore layer in the skin (39). Chromophore contents were given in arbitrary units, calculating relative changes as percentage of chromophore content in control skin (39).

Casual Sebum

Casual sebum content may be evaluated photometrically using a Sebumeter (SM 810 PC, Courage & Khazaka, Cologne, Germany; Acaderm, Menlo Park, CA). This device works on the principle of photometry of a special quilted plastic strip that becomes transparent with fat absorption (a high lipid level causes a greater transparency of the plastic strip) (40). A mirror is fitted underneath the plastic strip and is connected with the housing by means of a pressured spring. Thus, casual lipid is collected from the skin with a constant pressure of 6 N for 30 s. The reflected beams are measured and transferred to a digital instrument.

Skin Surface Temperature

Skin surface temperature can be evaluated with standard temperature measuring devices and by thermography. Skin temperature is important in directly assessing the irritation response of skin and has many indirect effects on other parameters, including TEWL.

Thermography has previously been described in detail and is recognized as a method of producing high-resolution image profiles of skin surface temperature (23,41,42). A membrane consisting of cholesteric liquid crystals on a black light-absorbent material with high thermal conductivity is used to depict the temperature distribution over the test area (Flexitherm contact-thermographic equipment) (42). The crystals are between liquid and solid phase, and the color of the crystals depends upon the given temperature—i.e., the crystals reflect brown light at low temperatures and blue light at high temperatures in the temperature range 23.5–29.1°C; changes is color can indicate temperature differences of 0.6°C (41). The thermography membrane is mounted in a frame that can be insufflated to aid contouring with the skin surface.

Mechanical Properties

The Dermal Torque Meter, Dermaflex, Gas-Bearing Electrodynamometer, and Cutometer (Acaderm, Menlo Park, CA) belong to a general class of instruments that measure skin phenomena such as skin angular motion, elasticity, and other skin rheology (27). These instruments are used to evaluate such products as skin moisturizers, surfactants, and corticosteroids by using skin suction, torsion, indentation, and elevation to contort the skin and then measure its return to its original state (23). They are particularly useful in measuring wound healing, solar aging effects from photodamage, and other dermatological skin conditions (23).

Skin Thickness

Skin thickness is most commonly measured using the noninvasive ultrasound technique. Ultrasound is an important tool in experimental dermatology, as data on skin thickness reveals the degree to which skin has been affected by such conditions as psoriasis and scleroderma (43). Ultrasound is also a viable alternative to skinfold calipers and is to be preferred when measuring the thickness of uncompressed subcutaneous adipose tissue (44).

In conclusion, these techniques provide powerful tools in understanding the skin of the consumer. Yet we emphasize that the consumer is the tool by which the techniques are evaluated—in appropriate clinical trials.

REFERENCES

1. Allen AM. Design methodology in trials of topical drugs.
2. Allen AM. Clinical trials in dermatology: Part 1. Experimental design. Int J Dermatol 1978; 17:42–51.
3. Fleiss JL. Statistical Methods for Rates and Proportions. New York: Wiley, 1973:30–194.
4. Marples RR, Kligman AM. Limitations of paired comparisons of topical drugs. Br J Dermatol 1973; 88:61–67.
5. Finney DJ. Experimental Design and Its Statistical Basis. Chicago: University of Chicago Press, 1955:3–58.
6. Allen AM. Clinical trials in dermatology: Part 2. Numbers of patients required. Int J Dermatol 1978; 17:194–203.
7. Tversky A, Kahneman D. Belief in the law of small numbers. Psychol Bull 1971; 76:105–110.
8. Allen AM. Clinical trials in dermatology: Part 3. Measuring responses to treatment. Int J Dermatol 1980; 19:1–6.
9. Grice K, Sattra H, Sharrat M, Baker H. Skin temperature and transepidermal water loss. J Invest Dermatol 1971; 57:108–110.
10. Dugard PH. Skin permeability theory and relation to measurements of percutaneous absorption in toxicology. In: Marzulli FN, Maibach HI eds. Dermatotoxicology and Pharmacology. New York: Wiley, 1977: 525–550.
11. Hassing JH, Nater JP, Bleumink E. Irritance of low concentrations. Dermatologica 1982; 154:314–321.
12. Maibach HI, Bronaugh R, Guy R, et al. Noninvasive techniques for determining skin function. In Drill VA, Lazar P, eds. Cutaneous Toxicity. New York: Raven Press, 1984:63–97.
13. Berardesca E, Maibach HI: Racial differences is sodium lauryl sulphate induced cutaneous irritation: Black and white. Contact Derm 1988; 18:65–70.
14. Berardesca E, Maibach HI. The effect of non visible irritation on the water holding capacity of the stratum corneum (abstr). Presented at the First European Symposium on Contact Dermtitis. Heidelberg, West Germany, May 27–29, 1988.

15. Serup J, Straberg B. Differentiation of irritant and allergic reactions by transepidermal water loss. Contact Derm 1987; 16:129–132.
16. Pinnagoda J, Turker RA, Agner T, Serup J. Guidelines for transepidermal water loss (TEWL) measurement. Contact Derm 1990; 22:164–178.
17. Wilhelm KP, Cua AB, Wolff HH, Maibach HI. Surfactant-induced stratum corneum hydration in vivo: Prediction of the irritation potential of anionic surfactants. J Invest Dermatol 1993; 101:310–315.
18. Berardesca E, Fideli D, Gabba P, et al. Ranking of surfactant skin irritancy in vivo in man using the plastic occlusion stress test. Contact Derm 1990; 23:1–5.
19. Blank HI, Moloney J, Emslie A, et al. The diffusion of water through the stratum corneum as a function of its water content. J Invest Dermatol 1984; 82:188–194.
20. T Agner, J Serup. Skin reactions to irritants assessed by noninvasive bioengineering methods. Contact Derm 1989; 20:352–359.
21. Tagami H, et al. Evaluation of the hydration state of the stratum corneum in vivo by electrical measurement. J Invest Dermatol 1980; 75:500–507.
22. Blichmann C, Serup J. Assessment of skin moisture: Measurement of electrical conductance, electrical capacitance and transepidermal water loss. Acta Derm Venereol 1988; 68:284–290.
23. Wild JE. Biophysical instrumentation used to evaluate personal care products. Cosmet Toiletr 1993; 108:71–74.
24. Tagami H, et al. Water sorption-desorption test of the skin in vivo for functional assessment of the stratum corneum. J Invest Dermatol 1982; 78:425–428.
25. Jacques SL, Maibach HI, Suskind C. Water content in the stratum corneum measured by a focused microwave probe: Normal and psoriatic. Bioeng Skin 1981; 3:118–119.
26. Simon I, Emslie AG, Apt CM, et al. Determination in vivo of water concentration profile in human stratum corneum by a photoacoustic method. In: Marks R, Payne PA, eds. Bioengineering and the Skin. Lancaster, England: MTP Press, 1981:187–195.
27. Kajs TM, Garstein V. Review of the instrumental assessment of skin: Effects of cleansing products. J Soc Cosmet Chem 1991; 42:249–271.
28. Wicket RR, Trobaugh CM. Personal care products: Effects on skin surface pH. Cosmet Toiletr 1990; 105(7):41–46.
29. Wilhelm KP, Maibach HI. Skin color reflectance measurements for objective quantification of erythema in man. J Am Acad Dermatol 1989; 21:1306–1308.

30. Nilsson EG, Tenland T, Oberg PA. Evaluation of a laser Doppler flowmeter for measurement of tissue blood flow. IEEE Trans Biomed Eng 1980; 27:597–604.

31. Berardesca E, Maibach HI. Cutaneous reactive hyperemia: Racial differences induced by corticoid application. Br J Dermatol 1989; 120: 787–794.

32. Nilsson GE, Otto U, Wahlberg JE. Assessment of skin irritancy in man by laser Doppler flowmetry. Contact Derm 1982; 8:401–406.

33. Staberg B, Klemp P, Serup J. Patch test response evaluated by cutaneous blood flow measurements. Arch Dermatol 1984; 120:741–743.

34. Staberg B, Serup J. Allergic and irritant reactions evaluated by laser Doppler flowmetry. Contact Derm 1988; 18:40–45.

35. Dorogi PL, Zielinski M. Assessment of skin conditions using profilometry. Cosmet Toiletr 1989; 104(3):39–44.

36. Grove GL, Grove MJ. Objective methods for assessing skin surface topography noninvasively. In Leveque JL, ed. Cutaneous Investigation in Health and Disease: Noninvasive Methods and Instrumentation. New York: Marcel Dekker, 1989:1–32.

37. Groh DG, Mills OH, Kligman AM. Quantitative assessment of cyanoacrylate follicular biopsies by image analysis. J Soc Cosmet Chem 1992; 43:101–112.

38. Bjerring P, Anderson PH. Skin reflectance spectrophotometry. Photodermatology 1987; 4:167–171.

39. Andersen PH, Bjerring P. Noninvasive computerized analysis of skin chromophores in vivo by reflectance spectroscopy. Photodermatol Photoimmunol Photomed 1990; 7:249–257.

40. Courage W, Khazaka G. Instruction Manual for Sebumeter SM 410. Cologne: Federal Republic of Germany, 1988.

41. Agner T, Serup J. Contact thermography for assessment of skin damage due to experimental irritants. Acta Dermatol Venereol (Stockh) 1988; 68:192–195.

42. Ring EFJ. Skin temperature measurement. Bioeng Skin 1986; 2:15–30.

43. Sondergaard J, Serup J, Tikjob G. Ultrasonic A and B scanning in clinical and experimental dermatology. Acta Dermatol Venereol 1986; 66 (suppl 120):75–82.

44. Jones PRM, Davies PSW, Norgan NG. Ultrasonic measurements of subcutaneous adipose tissue thickness in man. Am J Phys Anthropol 1986; 71:359–363.

4
Methods for Evaluating the Effects of Therapies and Extrinsic Factors on Acne

Otto H. Mills, Jr.
Hill Top Research, Inc., East Brunswick, New Jersey

INTRODUCTION

Acne vulgaris, not unlike other abnormalities of the skin, has both intrinsic and extrinsic factors that influence the pathology. Traditionally cited intrinsic factors include (1) excessive sebum, (2) hyperkeratinization of the sebaceous follicle, and (3) rupture of the follicular epithelium, which leads to the appearance of the inflammatory lesions. The microorganism *Propionibacterium acnes* has also been said to play a role in the inflammatory phase of this disease (1).

Extrinsic factors that affect the acne-prone population include: friction (2) and certain topically applied or orally administered substances (3). Sunlight may also have a negative influence on acne-prone skin, even though it appears to dry and camouflage the erythema (4,5).

The following brief chapter addresses the intrinsic and extrinsic factors associated with acne vulgaris and outlines some of the approaches for human testing in these areas. Intrinsic factors can be addressed by both clinical trials and model systems. Extrinsic factors can be tested for to some extent and must also be watched for when testing antiacne agents for efficacy. As requirements may differ, particularly depending on the goals of your study, it is, of course, always best to take the pulse of the appropriate regulatory agency.

EVALUATING THE EFFECT OF THERAPY ON INTRINSIC FACTORS

Doing efficacy trials in acne, like the condition itself, would seem obvious, with the elements on the surface. However, not surprisingly to those familiar with the skin, the eye's view, connected to human confidence, can lead to a superficial view of the task at hand. One need only to consult the literature to see the number of agents that have been found to work or "cure" this common, chronic problem (6). A few authors have reported contradictory results or published a listing of agents that do not work in acne (7).

If one wishes to know whether an agent works in acne, one must recognize that the disease has two types of lesions: (1) noninflammatory (open and closed comedones) and (2) inflammatory (papules, pustules, and nodules). Using existing marketed agents as benchmarks, one can expect to see an impact on the inflammatory component of acne in a very short period of time (days to weeks). Based on existing products, the noninflammatory lesions will respond more slowly and may take weeks to months until a reduction is seen. Therefore, in general, a standard acne efficacy trial lasts 8–12 weeks and sometimes as long as 4 months, depending on the projected mechanism of action of the modality being tested. Two types of evaluations are done during this time period. One centers on the evaluator *characterizing* and *counting the clinically evident lesions* and the second is a *global evaluation*, which usually compares the overall appearance of the disease to the baseline line. Frequently protocols will also ask that prospective volunteer acne subjects not use any medicine topically for 2 weeks or orally for 4 weeks before the baseline evaluations.

In order to include the degree of acne that is being targeted with a product, one must set minimum and some maximum requirements. Most of the acne market would appear to be concentrated in the range of mild to moderately severe inflammatory acne vulgaris. Here, many times, protocols will indicate a minimum number of 10 to 15 noninflammatory lesions and 10 to 15 inflammatory lesions. Frequently subjects with cysts are excluded or maximum numbers of 1 to 5 may be used. Size should be determined for the latter, and maximum papule size should be set. Also, as lesion numbers can vary greatly, the setting of maximum numbers can better define a specific population, which may be helpful in future statistical analyses. Evaluations usually occur at baseline and then at weeks 1 or 2, 4, 8, and 12. Another evaluation scenario is baseline and then weeks 3, 6, and 12. The former design with a visit at the end of the first or second week is sometimes helpful in (1) reinforcing compliance, (2) gaining accurate information regarding any problems (clinical or symptomatic) or concerns with the use of the new medication, and (3) documenting early reductions in lesions.

Lesion characterization and counting requires practice and concentration. A good exercise to do before beginning the facial trial is to have the evaluator count lesions on 10 acne subjects, take a break, and in randomized fashion repeat the evaluation on the same 10 subjects. This exercise is a good experience for the evaluator and lets the potential sponsor know the variation to be expected if this evaluator participates. By doing this with all the evaluators in a potential multicenter trial, the sponsor can also project the variation from center to center.

Global evaluations are important because they indicate whether the agent under test is in fact having a clinically significant impact on the disease. Sometimes statistical analysis of lesion counts must be viewed carefully. For example, a 15% reduction in noninflammatory lesions may not be clinically significant to anyone—the evaluator, the acne subject, or an accompanying parent.

Conceptually, a global evaluation makes very good sense; however, the mechanism for doing such evaluations is sometimes confusing and does not fit the definition. Many protocols have global evaluation scales defined by percent improvement from baseline. If only lesion counts exist at baseline, the evaluator has been forced to consult these and the current visit's counts and draws a "global" evaluation based on these. Obviously, the quantitative counts are used twice and a true global evaluation may be lost in the process.

A number of ways have been used to get a global sense of improvement or lack of it. A standardized set of photographs has been developed by William Cunliffe (8), and these are used to grade lesions with the acne subject at each visit. Each photograph has a numerical value assigned to it (e.g., 0.75, 1.50, 3.00, 45.0, etc.). Thus one can use these good-quality photographs to judge global changes. This approach has some drawbacks. The main one may be that the photographs in the middle of the scale are not that different in some regards, whereas the differences between the mild and severe disease states are clear. However, this approach has strengths that recommend it, including its universal availability.

One can also use absolute global grades for each visit. For example, the evaluator uses a scale from zero to ten representing no acne to very severe acne. This grading again allows for a comparison of global change relative to each visit and more importantly to baseline.

If one wishes to truly judge global changes relative to baseline, one can take good standardized photographs (a left and right profile) and hold these up to the acne subject at each evaluation. The global evaluation is an important part of judging the efficacy of agents, because this type of evaluation points to clinical significance by capturing a total picture of the disease state that includes number, size, and intensity of lesion erythema.

Evaluating Safety Issues

Many acne medications have some inherent irritation level. Most acne subjects know this by experience. Recording the clinical and symptom changes is important in order to profile the potential new product. The early part of the clinical trial will frequently tell the story of how much irritation is experienced and how this may influence compliance in actual use if and when the product is marketed. As the skin tends to accommodate to mild signs and symptoms of irritation, an early clinical visit at 1 or 2 weeks after starting with an evaluation may be important to capturing this information. Clinical signs that can be evaluated include erythema, edema, dryness, and peeling. Symptom reports of burning, itching, stinging, and tightness may in some ways be even more important. A basic scale of none = 0, mild = 1, moderate = 2, and severe = 3 is fine, with more elaborate 10-point scales used less commonly.

Also, diaries that include a place for comments can pick up on early problems which may influence the patient's compliance. Perceptual questionnaires that are completed by the subjects independent of the evaluator can also be helpful in gathering data. These need to be reviewed before the subject leaves the site to ensure that a moderate or severe report is investigated to see if a true adverse reaction exists.

Marketing Questionnaires: Perception and Preference

Subjects' perceptions, of course, are important not only in areas of symptoms, as mentioned above, but also in the areas of likes an dislikes regarding the test product. Sensory questions that are well designed and organized to allow for meaningful analysis can provide information that may be helpful for differentiating products in the marketplace. Areas of interest are similar to those defined in sensory panel testing and can also be outlined prior the large clinical trial (9). Questionnaires filled out at the testing site will produce more usable data than those done at home, because someone is available to clarify questions and review the forms to be certain all information has been supplied.

Adjunctive Issues

Choice of a cleanser for an acne trial is critical. As the traditional format for product use is to wash and apply, the study sponsor must be aware that the detergent system in use before the product is applied greatly influences the signs and symptoms found in an acne trial, particularly those attributed to a product following immediate application. Sometimes the changes are transient and last for seconds to an hour or two and are capable of influencing usage patterns. Standardizing the cleansing system and supplying the study participants with the appropriate number of units is one good approach. One can, however, allow the study subjects to continue using the cleanser they have used before entering the trial and note the product on the admission form.

Sunscreen or sun block issues must be addressed in protocol design. Subjects should be instructed to avoid sun exposure, and one can exclude these people if the signs and symptoms of such are evident. Some subjects use a sunscreen or sun block regularly, and here again the issue of standardization enters.

Cosmetic products such as covers, foundations, powders, etc., are part of some female subjects' daily lives. One can recruit females who do not use facial cosmetics. Obviously, this increases time and costs. One can prohibit usage on the days of evaluation when it is important not for on the evaluation but for any photography. The use of medicated shampoo for dandruff or the perception of dandruff is not uncommon in this population. If subjects are using such products, it is difficult to understand how these will influence their acne because a clinical trial starts. Starting any new medicated shampoo must to be considered by the evaluator.

Medications taken orally or applied to other than the study sites must be considered case by case using guidelines set up in the protocol.

Seasonal Considerations

Not unlike other common chronic skin conditions, acne appears to undergo an annual cycle of severity. It is important not to start a study during the summer months, when the disease often appears to improve. Also, a number of cases may worsen (10), and this may be related to sun exposure (12). In our experience, the best time to do acne trials is fall through spring.

EVALUATING EXTRINSIC FACTORS

Friction

Static or kinetic friction on acne-prone skin frequently leads to an exacerbation of the disease where the friction occurs (2). The pattern of the exacerbation will often identify the source of friction — hat, football shoes, pads, etc. The exception to this may, in fact, be the most troublesome. Acne subjects may be washing or cleansing their faces with a significant amount of friction whether they use their hands, a washcloth, or a scrubbing device. A reduction of friction with gentle washing should be stressed in acne evaluation in order to avoid negative effects.

"Nonacnegenic" Claims

For us, the term *acnegenic* comprises two phenomena: ingestion or application of agents (3). *Acnegenicity* refers to the creation of noninflammatory lesions (comedogenic) or inflammatory lesions (pustulogenic or follicular irritation). At the current time, these two phenomena require different testing approaches. In order to assess comedogenicity (generation of a comedone) one must use a test system that detects an increase of comedones accurately. Although the literature describes an animal model (13,14), we favor the use of a human back model (15,16). This model has been shown to identify positive agents by expert evaluation and image analysis (17).

Basically, the assay is 1 month in duration and uses the follicular biopsy as the sampling technique for assessing microcomedone formation on the back. Screening for subjects with a tendency to form these lesions is key. Also, one is able to include positive and negative controls. The latter are always included, as all final comparisons are made to the individual's background ability to form impactions.

The phenomenon of comedogenicity centers on the formation of a hyperkeratinosis impaction in the sebaceous follicle. This formation, in our hands, can be detected in humans in a reasonable amount of time only by using the follicular biopsy. Clinically evident lesions may not appear for months or years.

Pustulogenicity or irritant folliculitis appears to be detectible on the face in a period of a few weeks. This phenomenon may well represent the follicular epithelium's response to an agent or combination of agents. Probably the best way to screen agents without using positive controls is to do inflammatory lesion counts and global evaluations for 3 to 4 weeks in a susceptible population.

As the goal in acnegenic testing is to devise and validate a facial assay capable of screening for both comedogenicity and pustulogenicity, we continue to explore ways to do this (18,19).

SUMMARY

Acne vulgaris is a common, chronic problem that can extend well into adulthood. Recent work done by our group in collaboration with others found that 70% of postadolescent women had some acne.(20) This figure, along with the more common numbers for this disease in adolescence, makes the problem a very significant one. Much remains to be done to fully detail the pathology of the disease and many opportunities exist for personal care, over-the-counter, and prescription companies to introduce adjunct and therapeutic agents into this area.

REFERENCES

1. Leyden JJ, Nordstrom KM, McGinley KJ. Cutaneous microbiology. In: Goldsmith LA, ed. Physiology, Biochemistry, and Molecular Biology of the Skin. Vol II, pp. 1403–1424.
2. Mills OH, Kligman AM. Acne mechanica. Arch Dermatol 1975; 111:481–483.
3. Mills OH, Berger RS. Defining the susceptibility of acne-prone and sensitive skin populations to extrinsic factors. Dermatol Clin 1991; 9:93–98.
4. Mills OH, Kligman AM. Ultraviolet phototherapy and photochemotherapy of acne vulgaris. Arch Dermatol 1978; 114:221–223.
5. Mills OH, Porte M, Kligman AM. Enhancement of comedogenic substances by ultraviolet radiation. Br J Dermatol 1978; 98:145–150.
6. Frank SB. Acne Vulgaris. Springfield, IL: Charles C Thomas, 1971.
7. Mills OH, Kligman AM. Drugs that are ineffective in the treatment of acne vulgaris. Br J Dermatol 1983; 108:371–374.
8. Burke BM, Cunliffe WJ. The assessment of acne vulgaris the Leeds technique. Br J Dermatol 1984; 3:83–92.
9. Aust LB, Oddo LP, Wild JE, et al. The descriptive analysis of skin care products by a trained panel of judges. J Soc Cosmet Chem 1987; 38:443–449.
10. Cunliffe WJ, Cotterill JA. The Acnes. Vol 6. Philadelphia: Saunders, 1975:15.
11. Kligman AM, Mills OH. Sunlight and acne. J Dermatol 1978; 5:47–49.
12. Mills OH, Kligman AM. Acne aestivalis. Arch Dermatol 1975; 111:891–892.
13. Kligman AM, Katz AG. Pathogenesis of acne vulgaris. Arch Dermatol 1968; 98:53–57.
14. Kligman AM, Mills OH. Acne cosmetica. Arch Dermatol 1972; 106:843–850.
15. Mills OH, Kligman AM. A human model for assessing comedogenic substances. Arch Dermatol 1982; 118:903–905.
16. Mills OH, Berger RS, Dammers KS, Bowman JP. Assessing Comedogenicity. In: Waggoner WC, ed. Current and Future Trends in Clinical Safety and Efficacy Testing of Cosmetics. New York: Marcel Dekker, 1989:83–91.

17. Groh DG, Mills OH, Kligman AM. The quantitative assessment of cyanoacrylate follicular biopsies by image analysis. Soc Cosmet Chem 1992; 43:101–102.
18. Ayres JC, Mills, OH, Lyssikatos JC, et al. Assessment of a New Method for Determining Acnegenic Potential of Topically Applied Material on Human Subjects. Proceedings of the 17th Annual IFSCC International Congress Vol. 2. Yokohama, Japan, 1992:889–912.
19. Groh DG, Ayres JC, Mills OH, Kligman AM. Human Comedogenicity Testing: Correlation Between Image Analysis Evaluation and Expert Grader Scores. Proceedings of the 17th Annual IFSCC International Congress, Vol. 2. Yokohama, Japan, 1992:106.
20. Mills OH, Kligman AM, Hart J, et al. An Epidemiological Study of Acne in Post adolescent Females. 18th World Congress of Dermatology: Progress and Perspectives. New York, June 12–18, 1992:153A.

5
Skin Conditioning Benefits of Moisturizing Products

Kenneth D. Marenus
Estee Lauder Research Laboratories, Melville, New York

INTRODUCTION

Cosmetic products are broadly defined as agents that can improve the skin's appearance, and moisturizers fall within this definition. The ability of modern moisturizers to improve skin condition and appearance is widely recognized, especially since the effects of such products can be perceived by the consumer and measured using a variety of clinical techniques. In this chapter the potential benefits of moisturizers for both skin dryness and photoaging are discussed, as well as the means by which these benefits may be measured.

The term *moisturizer* is used to define a broad range of products. The best of these are designed to improve skin appearance by improving skin condition. Two basic skin conditions are usually in need of improvement: (1) "dry skin," which may be related to a variety of factors including defective barrier lipid synthesis, incomplete differentiation, and chronic low level inflammation, and (2) the cumulative effects of photoaging, which ultimately lead to the formation of lines, wrinkles, and a loss of elasticity.

Modern moisturizers can achieve improvements in both of these skin conditions in many ways. They can help optimize the differentiation process in the epidermis, thereby assuring timely production of appropriate lipids, establishment of an optimal barrier, and subsequent reduction of the "dry skin" condition. In turn, an optimally functional barrier contributes to a reduction in chronic low-level inflammatory responses. Low-level chronic inflammation can cause increases in the epidermal turnover rate and, as such, also contributes to the dry skin condition. In addition, there is now widespread agreement that chronic low-level inflammation can contribute to premature aging of the skin. Thus, modern moisturizers often contain soothing agents to reduce chronic low-level irritation.

In terms of photoaging, these products can effectively stimulate activity within the skin that results in distinct and perceivable improvements in surface topography as well as adjunct increases in skin firmness and elasticity. By improving the condition of the barrier and assuring optimal hydration, specific benefits in reducing small lines and wrinkles can be perceived and measured. What ultimately results from use of a modern moisturizer is smoother, softer, and firmer skin that feels more comfortable. Further, this skin is less responsive to low levels of irritants due to its improved barrier function and thus is better able to resist many of the factors that ultimately contribute to a less attractive appearance.

CLAIMS FOR MOISTURIZERS

As time progresses, challenges of product performance claims are increasing. Inquiries now come from all over the globe. They come not just from public health agencies but also from consumer protection groups, advertising watchdogs, and even small municipal governments. Claims for modern skin care products must be supported by data obtained from objective biophysical tests combined with results of well designed consumer perception studies. Taken together, these form the basis for claim support that can be used to satisfy the numerous inquiries arising when a new skin care product is marketed.

Claims concerning improvements in skin condition resulting from use of a moisturizer are usually couched in marketing phrases such as "Improves the appearance of lines and wrinkles," "Improves firmness and elasticity," "Softens and moisturizes," or "Reduces surface flakiness." Whatever the phrase, it is important for today's marketer to realize that that such claims may be challenged. Thus, to avoid regulatory or civil action, it is important to have support for each product claim. Further, new rules in the European Union require claim support information in the "product dossier" (1) for all new and existing cosmetic products.

Product claims come in two basic forms, comparative and preemptive. Comparative claims such as "Ours is better than theirs" are usually short-lived, unconvincing, and can result in good deal of pesky litigation. Preemptive claims, on the other hand, describe the benefits of a particular product and attempt to convince the consumer to purchase on the basis of a combination of clinical test data and testimonials.

Of the preemptive claims, the two most common are ingredient claims and performance claims. Ingredient claims are just that: "Contains botanical extract X," or "With special bioprotein mix," etc. Supporting such claims is relatively easy; either the product contains the promoted materials or it does not. If it does, an ingredient listing and simple explanation will usually suffice as claim support. It is important to note that if particular ingredients are featured in advertising, it is best that they have a proven function in the product.

Performance claims are an entirely different matter. Over the past several years a variety of objective biophysical methods have evolved that can effectively demonstrate changes in skin attributes such as firmness, elasticity, smoothness, and softness resulting from product use. Although the methods for such biophysical evaluations have not become standardized, application of basic scientific principles can create a reasonable basis for supporting product claims. In addition to biophysical measurements, effective support of performance claims may also involve clinical evaluation by trained technicians or dermatologists and consumer self-evaluation of the desired product benefits.

GENERAL EXPERIMENTAL DESIGN

Two basic experimental designs are frequently used in assessing moisturizer efficacy. The first involves the short-term effects of the moisturizer and its interaction with the skin surface while the second involves the subtle conditioning effects of longer-term use.

In short-term testing, panelists are usually examined during the course of a day after application of a test product at a defined dose. This type of testing is useful during the product development process to determine which of many possible prototypes might be the most desirable. It is important to note, however, that short-term testing is limited in its ability to discern the more subtle benefits that can be derived from moisturizers.

For claim support and final product testing, longer term home-use studies are recommended. These blinded studies usually involve two groups, treated and control, using the product at home for a period of 8 to 12 weeks. Treatment conditions are assigned to each panelist from the pool of panelists at random. In most cases it is good to have an untreated control group to account for anomalies in the weather that might affect the final results. The size of each group should be determined on a statistical basis, allowing for enough panelists in each group to assess significant changes (or not) at a confidence level of 95% or better.

Prior to the start of the study, panelists fitting the description of the end users of the particular type of product are selected. It is important to carefully define both inclusion and exclusion criteria for each study. These could be individuals with skin that is dry, oily, sensitive, or any other particular desirable parameter. Once panelists are recruited, the entire testing procedure is described to each panelist. They are then asked to sign an informed consent statement. Following admission to the study, each panelist also completes an extensive background questionnaire that inquires about family history, personal experience with respect to allergy and skin sensitivity, and current habits relating to grooming and use of toiletry products.

At the beginning of the study, prior to any product use, baseline clinical and instrumental measurements are made for each skin characteristic or treatment attribute that will be followed during the course of the study. These often involve measurements of firmness, surface topography, and smoothness. These are then repeated at regular intervals during the study, which are established in the initial study protocol, usually following 4 and 8 weeks of product use. A post-treatment "regression visit" is often scheduled at least 2 weeks after the last treatment to allow for assessment of residual effects of product use.

MOISTURIZERS AND DRY SKIN

Dry skin, although not a medical disease, is an uncomfortable condition that affects many people, especially in the winter. Characteristics of dry skin include flaking, itching, scaling, cracking, tightness, redness, and occasionally bleeding. The causes of dry skin, although not completely defined, are centered around four issues: lack of water in the stratum corneum, overactive epidermal turnover, inappropriate lipid synthesis, and barrier damage. All of these are interrelated; thus, the primary goal of any moisturizer is to reestablish optimal barrier function. If this is accomplished, it results in a reduction of the inflammatory response, with consequential reduction in epidermal turnover rate. This means that dry, flaky, cracked, and tight skin will be improved along with positive changes in appearance.

Currently, there are four key means by which to measure improvement in skin condition. First, there is measurement of transepidermal water loss, which can indicate the condition of the barrier. Next, there are measurements of surface conductivity, which yield information concerning the water content in the stratum corneum. Third is skin surface extensibility, which can be measured as an indication of degree of stratum corneum hydration. Finally, there is measurement of stratum corneum desquamation amounts and patterns removed by sticky tapes as an indication of the overall state of turnover and optimization of the differentiation process.

Barrier Condition and Transepidermal Water Loss

Instrumentation for measuring rates of transepidermal water loss has been available for several years. The measurement involves assessing the amount of water localized at two distinct points in space above the skin at the same time. Since the measurement of water content is made with reference to the skin surface at two distinct points simultaneously, a determination of the rate of water loss can be made.

The most common means of measuring transepidermal water loss has been by taking static measurements from individuals who have equilibrated in an environmental chamber. In fact, guidelines have been published as to how to make such measurements (2). Experimental evidence obtained in our own laboratory has shown that there is a better way to proceed.

The key reason for making transepidermal water loss measurements is to assess barrier condition. In order to make this assessment at the highest possible level of resolution, the barrier should be challenged during the course of the assessment. By taking tape strips of the stratum corneum and determining the number of strips required to reach a predetermined rate of water loss (15–18 $g/m^2/h$), this approach allows for more accurate discrimination of barrier condition than simple static measures. This point is illustrated in Figure 1. Here, after 7 days of exposure to a moisturizer containing a double amount of surfactant based emulsifiers, it is clear that there is no discernible difference in water loss between treated and untreated sites when measured statically.

However, if one then begins to strip the skin surface, it becomes clear that the skin treated with a moisturizer containing a higher load of emulsifiers is far less resistant to challenge than the untreated control sites. It is a dynamic measurement such as this which allows for determination of the more subtle effects of a cosmetic product on the barrier condition of the skin. This type of assessment should be made each time the panelists visit the lab during the course of a study.

Skin Water Content Measured by Conductivity

Once the effect of a product on barrier function has been determined, another important assessment of moisturization is measurement of water content in the stratum corneum. This is important, in that a fully hydrated stratum corneum provides for both a better barrier and a smoother surface. Measurement of the electrical properties of the skin at its surface can be made with a variety of instruments (3). The advantage of using this equipment is that, with it, it is relatively simple to generate data. This approach is thus convenient for assessments on large numbers of panelists and a wide variety of products. As in the case of other biophysical techniques, however, it is important that considerable care be taken in setting up the parameters of an experiment. For example, just as with water-loss measurements, panelists should equilibrate in an environmental chamber before conductivity measurements are made.

For short-term protocols, the contribution of the product to the measurement should also be considered. This determination can be made simply by following the conductivity of a thin film of the test material on the bench. For long-term protocols, panelists should generally be instructed not to apply the test material for 12 h prior to coming to the laboratory. This restriction should hold for all types of instrumental measurements. In the case of the short-term study, we are assessing the interaction of the product with the skin surface, while in the long-term study, the measurement is of changes in skin condition resulting from product use.

Stratum Corneum Hydration as Determined by Surface Extensibility Measurements

When the stratum corneum becomes desiccated, it acquires a certain stiffness. Conversely, when the stratum corneum is moisturized, its extensibility is increased. The increase in skin surface extensibility as measured using the gas-bearing electrodynanometer (4) is another good measure of moisturizer efficacy. A typical curve illustrating the ability of a moisturizing product to soften the skin surface is illustrated in Figure 2. This plot represents the area of the hysteresis loop created by the movement of the probe on the skin surface compared with time.

As can be seen from the curve, an initial softening effect is seen within the first few minutes. This is from the water in the product. As the water from the emulsion dries, there is a gradual return of the skin surface to its original state. The slower the return to its baseline value of extensibility, the more effective the moisturizer.

In terms of skin conditioning and softening benefits, the moisturizer should be evaluated during the course of an 8-week study in order to assess the ability of the product to plasticize the stratum corneum. This is another case where assessment of skin conditioning benefits are made; therefore, measurements are best taken 12 h after the last application of the test product.

Surface Squametry as a Diagnostic Tool

There is now a consistent supply of material designed for the specific purpose of stripping the stratum corneum; this has led to a small revo-

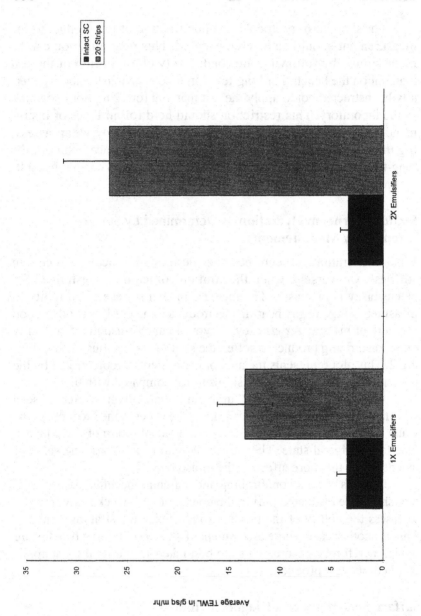

lution in the analysis of the skin surface. Using the D-Squame (5), it is now possible to analyze the stratum corneum both structurally and biochemically. This simple tool, modeled after adhesive tape, has made it possible to determine both patterns and amount of stratum corneum removed with each successive pull on the skin surface (6).

In addition to evaluation of both the amount and patterning of the stratum corneum, recent work indicates that it is possible to use stratum corneum removed in this manner to evaluate questions of a biochemical nature. Further, it is possible to use this simple tool together with analytical techniques, such as high-performance liquid chromatography (HPLC), to evaluate the partitioning and distribution of key effective ingredients from the moisturizer into the stratum corneum.

MOISTURIZERS AND PHOTOAGED SKIN

Aside from dryness, the other function ascribed to modern moisturizers is their ability to improve skin appearance with respect to smoothness and firmness. This change comes about because of the ability of these products to improve barrier function, thereby optimizing the differentiation process. In addition, there is impact in terms of increased water-holding capacity of the skin and new collagen synthesis. The result is a smoother and firmer skin.

As with skin dryness, there are now defined methods by which skin smoothness and firmness can be evaluated. Using the same exper-

---

Figure 1 Transepidermal water loss measurements of tape-stripped and unstripped stratum corneum after 7 days of treatment with moisturizers containing normal and double the amount of surfactant-based emulsifiers. It is clear from these results that the static measurement (unstripped) of transepidermal water loss does not resolve the more subtle changes in stratum corneum condition that have occurred in response to treatment with a moisturizer containing double the normal amount of emulsifiers. The dynamic measurement (tape-stripped) of barrier condition does allow for this comparison.

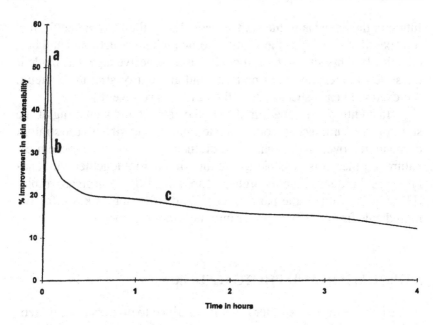

Figure 2 Effect of a typical moisturizer (oil in water emulsion) on skin surface extensibility as measured by the gas-bearing electrodynanometer. The initial increase (a) is due to the immediate delivery of water to the stratum
corneum from the water phase of the emulsion. A rather rapid decline in skin
surface extensibility also occurs as the water from the emulsion dries (b). The
residual effect (c) is due to the emoliency of the formula and its ability to provide a softening benefit that will last for several hours.

imental design described above, these measurements can be made in a
noninvasive manner and can provide objective data that help to support
product claims as well as to guide the product development process.

Skin Surface Replicas and Their Measurement

A key issue for a moisturizer is its ability to improve the appearance
of the skin surface by smoothing and reducing the appearance of fine
lines and wrinkles. Analysis of the changes in the skin surface requires
two things: first, an accurate means of reproducing surface topography

and, second, a reliable and objective means of quantifying surface changes.

The standard for reproducing surface topography in the industry is the silicone replica. These replicas are quickly and repeatedly made and can recreate skin surface detail to a resolution of better than 1.0 μm. When first made, the silicone replica is a negative impression of the skin surface. Some investigators prefer to make a positive casting from the silicone replica to produce a positive impression and exact reproduction of the skin surface.

Some debate still occurs as to whether a negative replica is adequate or whether a positive replica must be produced. For the most part, negative replicas will suffice for many applications. The balance to be considered is what is lost in making the positive versus what, if anything, is lost when the negative impression is analyzed.

Whatever the case, the replicas are usually analyzed via one of three techniques, physical profilometry, optical profilometry via digitized image analysis, and laser profilometry.

Physical profilometry was the one of first methods of replica analysis developed. In this approach, a counterbalanced stylus is drawn slowly across the surface of the replica. As the stylus is deflected by the contours of the surface, the variations are recorded by a variety of electronic instruments. The drawback of this technique is that it is limited in terms of the number of lines or scans that can be made across a single replica. Thus, in some cases, information is lost. Still, the technique is important historically in that it was the first adequate approach to this kind of analysis.

The second approach to replica assessment, optical profilometry, involves computer imaging techniques; that is, digitized image analysis. In this technique, the replica is illuminated from a fixed low angle (18–25). Lines, wrinkles, and other surface irregularities molded into the negative replica are then cast as shadows. This image is then captured, digitized, and imported into the analysis program. Here, threshold gray values are set according to established standards. The total area of the lines and their density is then determined. In addition, the standard deviation of the range of reflected light intensities (gray values) can also be assessed as an indication of surface uniformity on the replica.

Figure 3 illustrates the computer's view of images from replicas

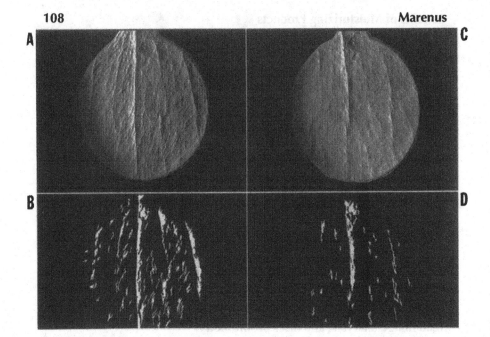

Figure 3   Digitized images of silicone skin surface replicas before and after 8 weeks of treatment with a modern moisturizer designed to improve skin surface topography. Frame A is a replica from an individual prior to use of the moisturizer; frame B is the digitized image of that replica. Frame C is the same region from the same panelist after 8 weeks of product use. Frame D is the corresponding digitized image from this replica. It is comparison of the areas of the digitized image in frames B and D that can yield a quantitative assessment of the improvement in skin surface topography resulting from use of the product.

taken before and after 8 weeks of treatment with a nourishing moisturizer. Here obvious visual differences in the replicas can be seen. These differences can easily be quantified with digitized image analysis. Further, it is important to note that differences such as these are also easily perceivable by the panelist and the consumer. As such, they are relevant for use in claim support efforts.

The third and latest approach to analyzing skin surface replicas involves laser profilometry. In this technique, a small laser spot is focused and scanned across the replica in both the X and the Y planes. Here, the laser determines the heights of peaks and the depths behind the ridges by an autofocusing technique. By using an arrangement of objective lenses, the focal point of the lens is adjusted until the incident and reflected beams are equivalent (7). This point is then compared to a fixed distance and the difference represents the wrinkle depth. Although this technique is relatively new, it may someday become the primary approach to replica analysis because of its potentially high resolution and versatility.

Skin Firmness and Elasticity

Two of the adjunct changes in skin behavior resulting from the photoaging process are losses of firmness and elasticity. Although structurally related to the condition of both collagen and elastin, these two parameters are functionally intertwined. Losses of firmness are usually accompanied by losses in elasticity. Firmness is the skin's ability to resist deformation, and elasticity is the rate at which a specific deformation is resolved. A variety of instruments currently exists to measure both of these functions. The most popular of these are the ballistometer, twistometer, and suction methods.

The ballistometer, based on the work of Tosti (8) and Hargens (9), is a light pendulum that is dropped from a fixed height onto the skin surface. The rebound of the pendulum is directly proportional to the firmness and elasticity of the skin surface. This measurement then can be used to reflect the degree of skin "age" in terms of firmness as well as the ability of a moisturizer to reverse the indication (10). Figure 4 illustrates the rebound curves of an individual before and after 8 weeks of treatment with a modern "nourishing" type of moisturizer.

The twistometer uses a different approach to assess changes in skin elasticity. This device employs a twisting motion to the skin surface for a short period of time. When the torque is released, the amount of time required for the skin to return to its resting position is determined (11). This device has been widely used to make measurements of both elasticity and extensibility. By comparing these two values, it is possible to evaluate a range of viscoelastic skin properties.

A few devices are commercially available that employ the suction

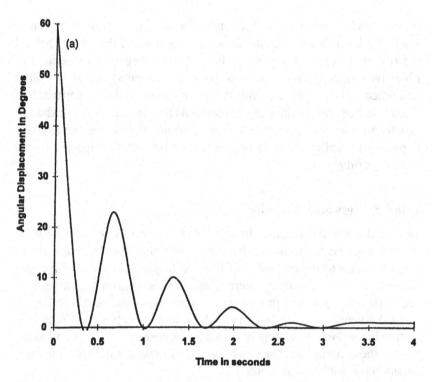

Figure 4 (a) The typical rebound pattern from an individual prior to the start of an 8-week moisturizer study. (b) The rebound pattern from the same site on that same individual after 8 weeks' use of a moisturizer designed to improve skin firmness. There are several ways to quantify changes in skin firmness, the most popular being the area under the first peak or the ratio of the heights of the first two peaks (H1/H2).

method for testing skin elasticity (12). With these devices, a gentle vacuum is applied to the skin surface. As the vacuum is applied, a small, sensitive gauge is displaced. The amount of displacement from the small fixed vacuum is a measure of skin extensibility. When the suction is released, the time required for the skin to return to its resting place is an indication of the skin's elasticity.

Each of these methods has its own advantages, disadvantages, and idiosyncracies. As with any other type of biophysical measurement,

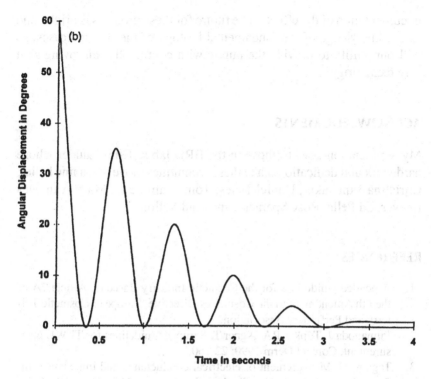

Figure 4 (Cont.)

however, the quality of the result is directly related to the exacting care of the operator.

SUMMARY

Modern moisturizers can do more than simply rehydrate the skin. In many cases these skin care products can provide distinct benefits in terms of skin condition and appearance. These benefits are achieved by applying the most modern techniques in product formulation combined with an understanding of skin biology. The benefits of these products can be demonstrated in clinical studies, which can provide objective documentation of the effects. The future for these products is bright, and as our knowledge of the fundamental biology of the skin increases, so will our ability to provide the public with continually improving skin care technology.

ACKNOWLEDGMENTS

My appreciation goes to those in the BRD lab at Estee Lauder whose hard work and dedication make these techniques move like a fine ballet: Christina Fthenakis, Daniel Maes, Tom Mammone, Mary Ann Mc-Keever, Ed Pelle, Rose Sparacio, and Paul Vallon.

REFERENCES

1. Amended Guidelines for the Cosmetic Industry Based on Article 7A of the 6th Amendment to the Cosmetics Directive, European Cosmetic Toiletry and Perfumery Association.
2. Pinnagoda J, Tupker RA, Agner T, Serup J. Guidelines for TEWL measurement. Contact Derm 1990; 22:160.
3. Tagami H. Measurement of electrical conductance and impedance. In: Serup J, Jemec G, eds. Handbook of Non-Invasive Methods and the Skin. Ann Arbor, MI: CRC Press, 1995:159.
4. Maes D, Short J, Turek BA, Reinstein JA. In-vivo measurement of skin softness using the gas bearing electrodynamometer. Int J Cosmet Sci 1983; 5:189–200.
5. Schatz H, Kligman AM, Manning S, Stoudemayer T. Quantification of dry (xerotic) skin by image analysis of scales removed by adhesive discs (D-Squames). J Soc Cosmet Chem 1993; 44:53–63.
6. Schatz H, Altmeyer PJ, Kligman AM. Dry skin and scaling evaluated by D-Squames and image analysis. In: Serup J, Jemec G, eds. Handbook of Non-Invasive Methods and The Skin. Ann Arbor, MI: CRC Press, 1995:153.
7. Efsen J, Hansen H, Christiansen S, Keiding J. Laser profilometry. In: Serup J, Jemec G, eds. in Handbook of Non-Invasive Methods and the Skin. Ann Arbor, MI: CRC Press, 1995:97.
8. Tosti A, Giovanni C, Fazzini M, Villardita S. A ballistometer for the study of plasto-elastic properties of skin, J Invest Dermatol 1977; 69:315.

9. Hargens, C.W., Ballistometry, In: Serup J, Jemec G, eds. Handbook of Non-Invasive Methods and the Skin. Ann Arbor, MI: CRC Press, 1995:359.
10. Fthenakis C, Maes D, Smith W. In-vivo assessment of skin elasticity using ballistometry. J Soc Cosmet Chem 1991; 42:211.
11. Leveque JL, Grove G, de Rigal J, et al. Biophysical characterization of dry facial skin. J Soc Cosmet Chem 1987; 38:171.
12. Serup J, Northeved A. Skin elasticity in psoriasis. In-vivo measurements of tensile distensibility, hysteresis and resilient distention with a new method: Comparison with skin thickness as measured with high frequency ultrasound. J Dermatol. 1985; 12:138.

6
Methods for Evaluating Sebum Removal

James D. Ayres
Amway Corporation, Ada, Michigan

INTRODUCTION

Sebum removal or reduction is often a major performance claim for cosmetic products. Given the wide variety of products acting on sebum levels, and the complex nature of the skin which secretes sebum, accurate measurement of product effects on sebum levels for claims documentation can be difficult. Effective claims documentation requires an understanding of the mode of product action, the process of sebum secretion, and the methods of sebum measurement.

Sebum is a secretion of the sebaceous glands, which are found in the epidermis throughout the body except on the palms, soles, and dorsa of the feet (1). These glands are holocrine and are found in highest concentrations in the face and scalp (2); consequently, sebum removal is an important performance characteristic of many skin care and cosmetic products. Sebum production is known to be largely controlled by endogenous hormones and varies from person to person depending on many factors (3). Levels are highest during the teenage years, falling off in women after menopause and remaining at relatively unchanged levels into old age in men (4). Sebum is composed of triglycerides, diglycerides, free fatty acids, wax esters, squalene, cholesterol, and cholesterol esters (5). The function of sebum can only be speculated upon. It

115

does have some mild bactericidal and antifungal properties, but it probably does not function in maintaining the skin's barrier function (6). Sebum is well known to protect the hair shaft cuticle from friction damage due to combing, brushing, and styling (7).

While the function of sebum in humans is somewhat unclear, excessive sebum on skin can have negative consequences. Excessive sebum leads to "oily skin," which feels greasy, looks patchy, and resists application of makeup. This condition is exacerbated by high humidity, where sweat can form an emulsion with sebum and increase the greasy feel. Kligman and Shelley suggest that the condition of oiliness in skin is actually dependent upon water and sweating (8). Additionally, excess sebum can contribute to packing of horny cells at the follicle surface, leading to an occlusive plug or comedone. Given these effects, it is no wonder that products designed for the "treatment" of oily skin represent a significant portion of the personal care product market.

Products for Oily Skin

Products for cosmetic treatment of oily skin typically are designed to emulsify or absorb excess sebum and facilitate removal from the skin surface. Cleansers that are either detergent- or soap-based emulsify the oils on the skin and facilitate rinsing and physical removal. Hydroalcoholic toners and lotions usually function to solubilize the sebum for easy removal with a tissue or cotton pad. Masks containing oil-absorbing materials can be very effective in pulling sebum from pores and the skin surface. Foundations for oily skin are designed to have a matte finish to decrease shininess and sometimes contain ingredients to counteract the effects of sebum on pigment shade. It is important to note that typical cosmetic products claim to remove sebum from the pores and skin surface or to ameliorate its effects. Products that claim to control actual sebum production by the sebaceous gland run the risk of being considered drugs.

METHODS FOR ASSESSING SEBUM OUTPUT AND REMOVAL

The performance claims for oily skin products usually revolve around the removal or absorption of surface sebum or reduction in surface levels over time. As such, it is critical to the cosmetic chemist to have scientific, reliable, and relevant methods for measuring product effects on sebum levels. Following are guidelines for documenting claims for sebum removal or reduction.

Experimental Design

Studies that measure sebum output or sebum removal from the skin are usually conducted on the forehead. This area is preferable for several reasons. First, the forehead offers a test area that is devoid of large terminal hairs, which may interfere with tape adhesion. Second, the forehead is a site of high sebum production with little interference by epidermal lipids (5,9,10). Third, this area can be readily studied on female subjects with minimal interference with normal makeup routines. Finally, due to bilateral symmetry, the forehead offers an ideal site for side-by-side testing of products and controls. Of course, the symmetry is not exact, and it is well known that many sebaceous glands can be in a quiescent state (11,12). It is also important to consider vertical variation in the pattern of sebaceous glands on the forehead. Sebum glands are in highest concentration on the forehead in the classic "t-zone" pattern. Consequently, paired sample sites must be contralateral, and effects of side-to-side variation must be minimized through the use of controls and an adequate sample size.

Sebum levels and output can vary dramatically, and the causes for differentiation must be taken into account when designing a study to measure sebum rates or removal. Piérard et al. found that sebum output varied with age or stage of life; sebum excretion rates were found to be differentiated into infantile, pubertal, acne, adult, and aging patterns (12). Menstrual cycle can also cause variation in sebum output in women with high sebum levels (13). Sebum output also varies seasonally, or, more accurately, with ambient temperature (14). Higher ambient temperature increases measured sebum excretion rates. Minor variations in the skin temperature can also affect sebum output (15), with changes on the order of 10% per 1°C variation being observed. It is not known whether these observed differences are due to changes in sebum viscosity and flow rates or effects upon the collection medium. Sebum excretion may also be affected by circadian rhythm (16).

These causes of variation must be considered in designing a study to document product effectiveness. Both ambient temperature and skin temperature must be taken into account. Studies should be run in controlled temperature and humidity settings if possible; both of these should be monitored and recorded throughout the study. Test subjects should be kept from excessive activity during the test period to minimize sweating. Assuming that the forehead is used, sample sites must be contralateral to minimize variation in the number of active glands sampled. Using a paired sample design on the forehead, analysis with paired sample t-test is appropriate.

Studies can be designed to measure the casual sebum level, cleansed sebum level, or sebum excretion rate. The casual level is the sebum that can be collected from skin which has not been recently cleansed or defatted. This measurement can be important in claims documentation studies for establishment of baseline before cleansing and in identifying oily-skinned subjects for testing. The cleansed sebum level is typically measured immediately after cleansing to assess the relative performance of a product. This approach can be used to compare formulas and products and to document sebum removal claims. Measurement of the sebum excretion rate can give an assessment of product effect on actual sebaceous gland activity.

Site Pretreatment

It is important to remove extant sebum from the test site at the beginning of the study. This ensures that the entire test site is at an equivalent baseline for sebum content. There are several methods to accomplish this, and the type and application of these can influence sebum measurement (17). Early work involved simply wiping the test site with dry gauze sponges (18). However, to ensure adequate sebum removal, it is advisable to utilize some sort of surfactant or solvent. Simply washing the forehead with soap and water can be adequate, but follow-up defatting with a hexane-saturated gauze pad can increase accuracy (19,20). Other researchers have used a gauze saturated with an ethanol solution of greater than 50% to wipe the test area (21,22). This may be preferred to using harsher solvents, as it removes superficial lipids without significantly altering the stratum corneum (17). For some studies, it may be advisable to use a cosmetic cleanser (23) or to have test subjects simply cleanse in their normal way before the test (24). After cleansing, the test site should be patted dry with tissue or cotton pad.

Panelist Selection

Prospective panelists should be screened for sebum level. This is typically achieved using the method that will be employed in the actual study. Data resulting from prescreening can be used for panelist selection and can provide baseline information. Test subjects for most product efficacy studies should typically be limited to those with oily skin. The higher sebum levels found in these subjects facilitates measurement of differences in product effect. Additionally, subjects should not be taking antiacne medication. When the forehead is used as the test site, contact with hands or hair can produce artifacts in sebum measurement. Consequently, it is advisable to use subjects who either have no bangs or are willing to keep them pulled back from the forehead for the duration of the study. Cunliffe and Taylor further recommend that subjects wash their hair on the evening prior to the test and apply no makeup or topical products thereafter (24). Subjects should also be advised to keep from touching, rubbing, or disturbing the test sites throughout the study (22).

Collection Methods

Many different methods have been developed to sample sebum from the skin surface. These include absorption into bentonite clay, transfer to ground glass, gravimetric paper absorption, and absorption to various tape substrates. These methods are detailed below.

Gravimetric Paper Absorption

This method was standardized by Strauss and Pochi in 1961 (18) and expanded in great detail by Cunliffe and Taylor in 1995 (24). Test areas on the forehead are marked off with adhesive tape, and the sites are defatted or cleansed thoroughly. Collections of sebum are made directly onto cigarette papers, which are pressed onto the test sites for periods of 1 to 3 h. After the collection period, the sebum content is assessed gravimetrically. The common method of analysis involves extraction of collected sebum with ether, then weighing the solute after ether evaporation. Simpler techniques involving direct measurement of the papers before and after collection have been described (25). The type of paper used for collection is critical to the accuracy of this method. Although many papers appear similar, they can vary greatly in their sebum absorptive capability (26). The original collection paper described in 1961 is no longer available, but new papers have been described. A special velin tissue, nonfluff paper was found by Cunliffe et al. in 1975 to give reproducible results (27). More recently, Cunliffe and Taylor recommended a paper from General Papers & Box Company (Severn Road, Treforest Industrial Estate, Poltypridd CF37 5SP England) as a suitable alternative (24). This method was designed to measure sebum excretion rates, but could be used to measure casual levels as well.

Clay Absorption

The clay absorption method was described by Downing et al. in 1982 (28). After cleansing, a film of an aqueous gel containing 15% bentonite clay and 0.2% carboxymethyl cellulose was applied to the skin and then covered with Dacron mesh. The film dried to an oil-absorbing layer, was left on for the prescribed time period, and then peeled off. This process was repeated several times. Collected lipids were extracted from the clay with ether, then measured by quantitative thin-layer chromatography. This method has the advantage of identifying the components of collected serum as well as quantifying them.

Lipometre and Sebumeter

Application of lipid to translucent surfaces alters their ability to transmit light. This fact has been the basis for the development of several methods to measure sebum. Schaefer first introduced a method for collecting sebum onto ground glass and measuring the resulting increase in light transmission (29). He found that the light transmission increase was proportional to the amount of surface lipid. This technique was improved with the creation of the Lipometre (L'Oreal, France), a simple, portable electronic device that can analyze the increased light transmission across a glass plate which results from deposition of lipid upon it (30). A ground glass plate is illuminated by a light-emitting diode which works at 650 nm and has a band width of 20 nm. The transmitted light is detected and amplified by a phototransistor. The light beam is 100% modulated at 3 kHz, and the receiving circuits are tuned to this frequency. Consequently, the apparatus is insensitive to environmental noise. The instrument calculates and displays the value of transmittance for each sample.

The Sebumeter (Schwarzhaupt Medizintechnik Germany) adsorbs the sebum onto a plastic film that has been rendered matte on one side. Under the foil is a mirror, which is connected to the housing in such a way as to exert a standard pressure on the area of skin being sampled. Once the head has been pressed to the test site for approximately 30 s, the head is inserted into the base and a lamp beams its light at a 45° angle onto the sebum-impregnated tape. The light is intensified by the mirror. The reflected rays are measured at 510 nm and the resulting number is displayed.

Both of these instruments are useful for measuring sebum. The Sebumeter is convenient, as there is no need to clean the sampling surface, and new tape is made available simply by the turn of a dial. Each Sebumeter tape can perform approximately 500 measurements before replacement is necessary. The Sebumeter is theoretically more accurate for a single measurement, having an error rate of about 10% as opposed to 15% for the Lipometre (31). As such, the Sebumeter is preferred for this single-measurement sampling. However, in cases where a site is measured several times in succession, the Lipometre is a more accurate method (31). Kligman et al. found the Lipometre to lack sensitivity at high and low levels of sebum excretion (20).

The Sebumeter represents an effective method of measuring the casual sebum level as well as sebum levels over time. Skin sites can be assessed before and at various times after product application. The resulting numerical data can be analyzed statistically to document product effects. Sebumeter data also lend themselves to graphical presentation for comparing products and for commercial purposes.

Sebutape and Instant Sebutape

The use of sebum-absorbent tape was first described by Nordstrom et al., in 1986 (19). The tape, known as Sebutape (CuDerm Corp., Dallas Texas), is made of a microporous, hydrophobic polymeric film composed of many tiny air cavities. The surface of the film is coated with a lipid-porous adhesive layer that enables the tape to adhere to the skin surface. The tape is applied to the skin test site for an optimal time of

1 h (9,21). Sebum on the skin surface is absorbed into the tape, displacing the air in the microcavities. As this occurs, the lipid-filled cavities become transparent to light. Through this process, the sebum output from each follicle forms a sharply defined clear spot, its size roughly corresponding to sebum volume (20). When the Sebutapes are placed on a black background, the sebum on the tape becomes clearly visible as black spots.

Sebutape can be analyzed in several ways to quantify sebum levels. First, collected sebum can be extracted from the tape, identified via thin layer chromatography, and quantified gravimetrically (20). However, because the sebum output from each follicle forms a clearly defined spot on the tape, several additional methods of analysis are possible. The area of the tape covered with sebum spots corresponds to the amount of sebum collected. This can be assessed visually for crude assessment. An improved method utilizing a densiometric device that measures the light transmitted through the tape sample has also been described (21). However, for exacting measurement, image analysis is recommended (9,20). Image analysis is based upon threshholding for gray level and then filtering of the image. The computer assigns a numerical value to each pixel of the image based upon brightness. This gray level can range in value from 0 to 256. These numerical values are then used to sort for quantification of various image objects per the definition set by the operator. Using this technique, the computer quantifies the sebum spots on each tape for number, area, and size. Data can be stored on disk and analyzed statistically.

Several factors must be considered when using image analysis for Sebutape quantification. With the equipment itself, a standard program must be used to ensure consistent processing for each image. Lighting must be controlled and standardized as well. It is also important to control storage of the tape samples and analyze them quickly and consistently. Changes can occur in the tapes that affect results. As sebum on the tape spreads. spots become larger and eventually fewer through lateral fusion of confluent sites. This can occur on tape samples within 60 min at room temperature (32) and will have a significant effect on results when numbers and activity of individual follicles is being measured (11). Storage in a freezer at −30°C can significantly slow this process, but it is still observed to occur within 7 days. Whitening of the sebum spots on the tape can also occur, probably due to crystallization (9,32). This problem can be eliminated through freezer storage (32). Because of these phenomena, Sebutapes should be evaluated either immediately or stored in frozen conditions and evaluated at a constant time interval from collection, preferably within 24 h (9).

Sebutapes are an ideal medium for assessment of sebum output over time. They have the added advantage of allowing assessment of individual follicles (20). Consequently, Sebutape offers a very exacting method for analysis of product effects on total sebum output and level, number of active follicles, and individual follicle activity, especially when measured with digital image analysis. Serup performed a comparison of Sebutape with the Sebumeter (21). He found that assessments from the two methods correlated but that the tape sometimes overestimated sebum output. The tape method was determined to be less reproducible over long sampling periods. This was postulated to be due to effects of occlusion and insulation while the tape is applied. However, precisely due to this effect and the interaction of sweat and temperature. Sebutape may be an ideal method for measurement of the cosmetic phenomenon of "oily skin" (21).

To facilitate good sebum collection from the skin surface. Sebutape is coated with an adhesive film. This ensures that the tape remain in contact with the skin throughout the test period. However, because of this, Sebutape is of limited use in measuring the casual sebum level with a short contact time (22). Instant Sebutape was developed to address this need. Instant Sebutape (CuDerm Corp., Dallas Texas) is based upon the same microporous film technology as Sebutape; however, it has no adhesive layer. Sebum levels are obtained by pressing the tape onto the skin test site for approximately 5 s. The Instant Sebutape surface is premounted upon a dark background, and sebum spots show immediately as dark spots.

The tapes are susceptible to error with variations in technique, and method of application must be tightly controlled to ensure accurate data. The pressure with which the tapes are applied to the skin is important. This can be standardized with the use of a Chatillon gauge (22). Application with a pressure of 350 g is recommended. Minor variations in the time of application can also affect results. Tapes are typically applied for 5 to 30 s, but a stopwatch should be used for consistency, and the application time should be standard for all samples in a study. Instant Sebu-tape is subject to the same "creep" phenomenon as Sebutape. They should be analyzed at a standard time from sampling, preferably within 24 h or stored in frozen conditions.

Instant Sebutape is a very convenient method for assessing the casual sebum level. The tapes can be analyzed either visually or with image analysis in the same manner as Sebutape. Using this technique, differences in sebum before and after product use can be clearly demon-strated and measured (Fig. 1). By quantifying the amount of sebum on the tapes with image analysis, effects of products on the casual sebum level over time can be assessed (Fig. 2).

Reece and Miller compared Instant Sebutape sebum measure-ments with those obtained with the Sebumeter (33). They found the methods to be comparable for most sebum levels, although the Sebume-ter measured quicker recovery after cleansing than Instant Sebutape. Whether the Instant Sebutape underestimates actual levels or Sebume-ter overestimates is open to speculation.

Scanning Electron Microscopy

Scanning electron microscopy is a well-known technique for obtaining very detailed, highly magnified pictures of surfaces. A scanning elec-tron microscope (SEM) utilizes an electron beam to sweep over a spec-imen that has been treated with a metal coating. The intensity of sec-ondary electrons generated at the point of impact on the sample is measured and used to generate the image. For studying skin sites, sili-cone replica technology is used to generate samples for SEM evaluation (34,35). Silicone replicas are typically prepared by catalyzing an RTV type silicone polysiloxane, spreading over the test site, and then peeling it off after polymerization has taken place and the material has set up. From this "negative" of the skin, a "positive" can be generated with any number of materials by creating a cast from the negative. Groh et al. rec-ommend pouring a molten polyethylene and ethylene/vinyl acetate copolymer blend over the negative, with centrifugation during cooling and setup (36). The resulting positive is a very detailed, permanent replica of the skin test site. Using this technique, individual pores can be examined before and after product application, and product perfor-

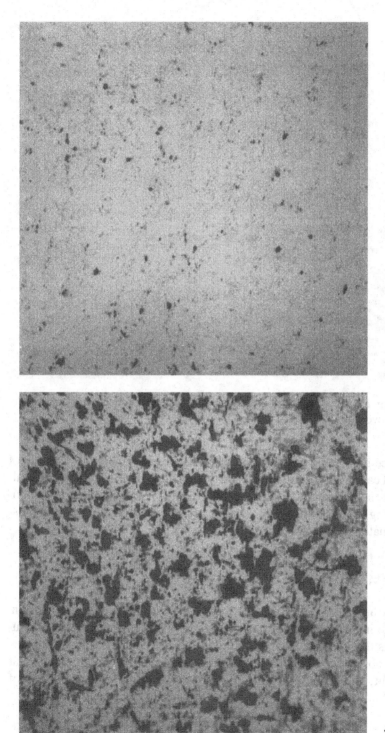

Figure 1 Instant Sebutape analysis of casual sebum level before and after use of a cosmetic cleanser.

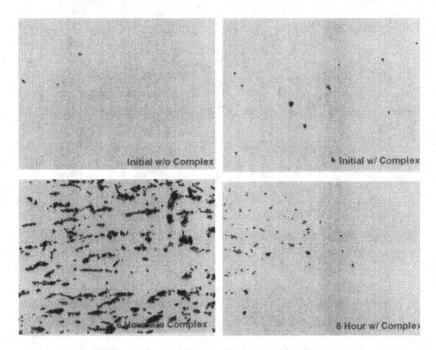

Figure 2 Image analysis measures the effect of products on casual sebum level over time.

mance in sebum removal can be readily observed (Fig. 3). The use of SEM can involve substantial cost as compared to other methods of evaluating sebum removal. However, this method has great utility in demonstrating product efficacy in sebum removal and generating merchandising materials.

SUMMARY

There are many well-proven methods to assess sebum levels on the skin. These include paper absorption and gravimetric measurement, instruments that measure changes in light transmission of a translucent surface, absorbent tapes that react with sebum to show individual follicular output, and scanning electron microscopy. When employing any of

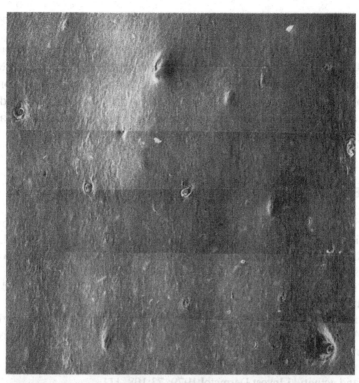

Figure 3   Scanning electron micrograph of sebum removal from pores by a clay mask.

these methods, variables such as panelist selection, test site preparation, product application, and sampling technique must be standardized to ensure accuracy. When this is done, these techniques represent important tools for the cosmetic chemist in assessing and documenting product effects on sebum removal, sebum absorption, and reduction in sebaceous gland activity.

REFERENCES

1. Strauss JS, Pochi PE. Histology, histochemistry, and electron microscopy of sebaceous glands in man. In: Gans O, Steigleder GK, eds. Handbuch der Haut-und Geschlechtskrankheiten; Normale und Pathologische Anatomie der Haut I. Berlin:Springer-Verlag, 1968:184–223.

2. Wilkinson JB, Moore RJ, eds. Harry's Cosmeticology, 7th ed. New York: Chemical Publishing, 1982.

3. Strauss JS, Downing DT, Ebling FJ. Sebaceous Glands. In: Goldsmith LA, ed. Biochemistry and Physiology of the Skin. New York: Oxford University Press, 1983:569–595.

4. Pouchi PE, Strass JS, Downing DT. Age related changes in sebaceous gland activity. J Invest Dermatol 1979; 73:108–111.

5. Greene RS, Downing, DT, Pochi PE., Strauss JS. Anatomical variation in the amount and composition of human skin surface lipids. J Invest Dermatol 1970; 54:240–247.

6. Kligman AM. The uses of sebum? In: Montagna W, Ellis RA, Silver AF, eds. Advances in Biology of Skin: The Sebaceous Glands. Oxford, England: Pergamon Press, 1963:110–124.

7. Agache PG. Seborrhea. In: Zviak C, ed. The Science of Hair Care. New York: Marcel Dekker, 1986:469–500.

8. Kligman AL, Shelley WB. Investigation on the biology of the human sebaceous gland. J Invest Dermatol 1958; 30:99–125.

9. el-Gammal C, el-Gammal S, Pagnoni A, Kligman AM. Sebum-absorbent tape in image analysis. In: Serup J, Jemee GBE, eds. Handbook of Non-Invasive Methods and the Skin. Boca Raton, FL: CRC Press, 1995:517–522.

10. Stewart ME, Downing DT, Strauss JS. Sebum secretion and sebaceous lipids. Dermatol Clin 1983; 1:335–344.

11. Piérard GE. Rate and topography of follicular sebum excretion. Dermatologica 1987; 175:280–283.

12. Piérard GE, Piérard-Franchimont C, Lê T, Lapiére C. Patterns of follicular sebum excretion rate during lifetime. Arch Dermatol Res 1987; 279:S104–S107.

13. Piérard-Franchimont C, Piérard GE, Kligman AL. Rhythm of sebum excretion during the menstrual cycle. Dermatologica 1991; 182:211–213.

14. Piérard-Franchimont C, Piérard GE, Kligman AL. Seasonal modulation of sebum excretion. Dematologica 1990; 181:21–22.

15. Cunliffe WJ, Burton JL, Shuster S. The effect of local temperature variations on the sebum excretion rate. Br J Dermatol 1970; 83:650–654.

16. Burton JL, Cunliffe WJ, Shuster S. Circadian rhythm in sebum excretion. Br J Dermatol 1970; 82:497–498.

17. Piérard GE. Relevance, comparison, and validation of techniques. In: Serup J, Jemee GBE, eds. Handbook of Non-Invasive Methods and the Skin. Boca Raton, FL: CRC Press, 1995:9–14.

18. Strauss JS, Pochi MD. The quantitative gravimetric determination of sebum production. J Invest Dermatol 1961; 36:293–298.

19. Nordstrom KM, Schmus HG, McGinley KJ, Leyden JJ. Measurement of sebum output using a lipid absorbent tape. J Invest Dermatol 1986; 87:260–263.

20. Kligman AL, Miller DL, McGinley KJ. Sebutape: A device for visualizing and measuring human sebaceous secretion. J Soc Cosmet Chem 1986; 37:369–374.

21. Serup J. Formation of oiliness and sebum output—Comparison of a lipid-absorbent and occlusive-tape method with photometry. Clin Exp Dermatol 1991; 16:258–263.

22. Groh DG, Ayres JD, Alexander RJ, Laskey KR. Ingredient specific "oil control" claims support using image analysis and Instant Sebutape. 10th International Symposium on Bioengineering and the Skin Cincinnati OH June 13–15, 1994.

23. Achtyes E, Scimeca J, Groh DG, Ayres JD. Possible effects of alpha-hydroxy acids on sebaceous gland activity. 54th Annual Meeting of the American Academy of Dermatology, Washington, DC, Feb 10–15, 1996.

24. Cunnliffe WJ, Taylor JP. Gravimetric technique for measuring sebum excretion rate. In: Serup J, Jemee, GBE, eds. Handbook of Non-Invasive Methods and the Skin. Boca Raton, FL: CRC Press, 1995:523–527.

25. Lookingbill DP, Cunliffe WJ. A direct gravimetric technique for measuring sebum excretion rate. Br J Dermatol 1986; 114:75–81.

26. Cunnliffe WJ, Shuster S. The rate of sebum excretion in man. Br J Dermatol 1969; 81:697–704.

27. Cunnliffe WJ, Williams SM, Tan SG. Sebum excretion rate investigations. Br J Dermatol 1975; 93:347.

28. Downing DT, Stranieri AM, Strauss JS. The effect of accumulated lipids on measurement of sebum secretion in human skin. J Invest Dermatol 1982; 79:226–228.

29. Schaefer H. The quantitative differentiation of sebum excretion using physical methods. J Soc Cosmet Chem 1973; 24:331–353.

30. Saint-Leger D, Berrebi C, Duboz C, Agache P. The Lipometre: An easy tool for rapid quantification of skin surface lipids (SSL) in man. Arch Dermatol Res 1979; 265:79–89.

31. Dikstein S, Zlotogorski A, Avriel E, et al. Comparison of the Sebumeter and the Lipometre. Bioeng Skin 1987; 3:197–207.

32. Pagnoni A, Kligman AL, el-Gammal S, et al. An improved procedure for quantitative analysis of sebum production using Sebutape. J Soc Cosmet *Chem* 1994; 45:221–225.

33. Reece BT, Miller DL. Comparison of two approaches for evaluating skin surface sebum. J Dermatol Clin Eval Soc 1992; 3:66.

34. Bernstein EO, Jones CB. Skin replication procedure for the scanning electron microscope. Science 1969; 166:252–253.

35. Garber CA, Nightingale CT. Characterizing cosmetic effects and skin morphology by scanning electron microscopy. J Soc Cosmet Chem 1976; 27:509–531.

36. Groh DG, Ayres JD, Maher P. Modified replication technique for preliminary laboratory evaluation of human skin surfaces. 16th Meeting of the International Federation of Societies of Cosmetic Chemists, New York Oct 8–11, 1990.

7

Methods for Claims Substantiation of Antiperspirants and Deodorants

John E. Wild and James P. Bowman
Hill Top Research, Inc., Miamiville, Ohio

Linda P. Oddo and Louise B. Aust
Hill Top Research, Inc., Scottsdale, Arizona

INTRODUCTION

Because specific claims may allow products to acquire market advantage, there are both legal and financial ramifications that must be taken into consideration. The manufacturer must be assured that the claims used to present and position products reflect the product's performance honestly and accurately. This chapter presents a discussion of the protocol characteristics that will provide reliable claim support evaluations for antiperspirant and/or deodorant products and limit the potential for challenge.

There has been rapid growth of the personal products and over-the-counter (OTC) drug and prescription drug industries in the past 35 years. With that rapid growth and the expansion of the market for these products, competition has increased. With increased competition, there has been an increase in product development, product improvement, and product niching. Antiperspirant products are no exception, using technology such as fragrance encapsulation for claims of longer-lasting deodorancy and product and packaging technology in the evolution of the various product forms such as gels and creams (1). These forms proved to be more esthetically pleasing than the chalky, residual solids. The clear gel imparts to the consumer the perception of purity, mildness, and less residue. However, with the advanced technology also came other issues that needed to be addressed, such as instability of formulations and effect of the product on sensitive skin (2).

Getting the consumer to purchase a product is a challenge the manufacturer must meet head on in order to be competitive in the marketplace. One way to get consumers to buy is to persuade them through advertising that a product can provide a desirable benefit. Once the product is purchased, the consumer provides a continuing evaluation and ultimately determines its success or failure.

TYPICAL STUDY DESIGN OBJECTIVES FOR CLAIMS SUPPORT TESTING

We have identified five typical study designs for claims support studies.

1. Substantiation of Product Definition.   Studies of product definition claims apply basic designs that select subjects having the characteristics for which the product is intended (i.e., they reduce underarm perspiration and/or body odor). These subjects are then placed on product use and observed for the change in the selected characteristic (i.e., perspiration or body odor). Figure 1 shows an example of antiperspirant efficacy results where the OTC standard for antiperspirant efficacy is achieved (3).

2. Substantiation of Product Performance Characteristics.   For performance characteristics testing, the procedure used for the product definition claim is specifically designed to test the identified performance characteristic (i.e., 24-h odor protection or 48-h wetness protection). Figure 2 shows an example of odor control performance of a deodorant product versus a placebo control test article. It supports a 24-h protection label claim.

3. Substantiation of Product Attribute Characteristics.   These study designs allow the sponsor to make claims about specific attributes that may characterize the product more distinctly to the consumer. Besides deodorancy, other skin-feel attributes such as mildness, ease of application, less residue, etc., can all be measured using appropriate study designs. Figure 3 shows a graph of consumer ratings of product attributes.

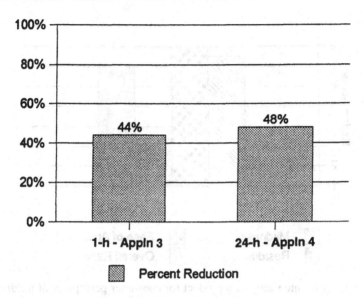

Figure 1   Antiperspirant product definition.

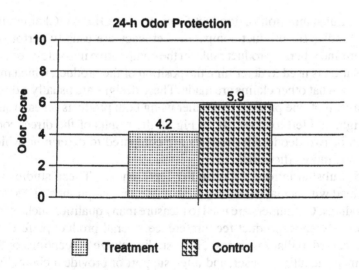

Figure 2   Product performance.

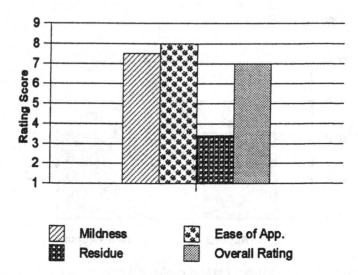

Figure 3 Attribute rating of a product for consumer perception of mildness, residue, ease of application, and overall acceptability. (Rating scale: 1 = dislike extremely, 5 = neither like nor dislike, 9 = like extremely.)

4. Substantiation of Product Comparative Efficacy Characteristics. Studies for product comparative efficacy are usually performed to determine where a product ranks in the competitive marketplace. This information is used to determine the position of the product, which may influence what other claims are made. These designs are usually a direct comparison of one product to another using both products on each subject (right and left axillae) (4). In Fig. 4, the results of the direct comparison of two deodorant products are compared to determine which product is more efficacious.

5. Substantiation of Consumer Preference. These studies are performed with actual consumer users of the antiperspirant and deodorant products. Consumers are used to measure many qualities such as fragrance preference, product-feel preference, overall product preference, and perceived axillary wetness. These studies utilize perceptions of the consumer, the ultimate user, and often support or provide a claim. Figure 5 illustrates such a study.

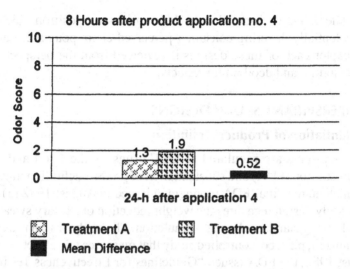

Figure 4 Comparative efficacy of two deodorant treatments. (Statistical analysis indicates a statistically significant difference favoring treatment A.)

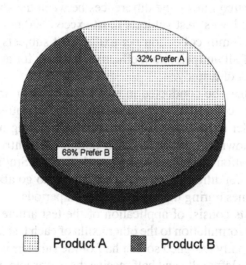

Figure 5 Product performance showing product B is preferred significantly more than product A.

These five designs all have different market position objectives and potentially differing consumer-perceived-claim potential. Later in the chapter each of these designs is reviewed from the perspective of antiperspirant and deodorant products.

ANTIPERSPIRANT STUDY DESIGNS

Substantiation of Product Definition

Antiperspirants are regulated as OTC drugs by the Food and Drug Administration (FDA) and substantiation of product definition must follow the tentative final FDA monograph issued in August 1982 (3). The basic study design measures the weight reduction of axillary sweat output following standard thermal stimulation. The study design is a blind, randomized, placebo-controlled study that measures this effect. Also, in August 1982, the FDA issued "Guidelines for Effectiveness Testing of OTC Antiperspirant Drug Products" (5). The FDA states that in order for a product to qualify as an effective antiperspirant product, it must meet the criteria established in these guidelines. A summary of these guidelines follows. Test subjects must be sufficiently representative of a variety of sweating rates—the differences between the sweat output of the highest and lowest test subjects must exceed 600 mg of sweat collected in one 20-min collection per axilla. Test subjects must abstain from the use of all antiperspirant effective products for at least 17 days prior to the start of a study.

A study may be conducted in either hot-room or ambient conditions. Hot-room conditions consist of 100°F and a relative humidity of 35–40% in order to thermally induce sweating. During hot-room testing, factors known to influence sweat output are controlled (i.e., air movement, position of the trunk, emotional stimuli). Studies conducted under ambient conditions allow the test subjects to go about their normal daily routines during the sweat-collection periods.

Treatments consist of application of the test article to one axilla and the control formulation to the other axilla of each test subject. Treatments are randomly assigned so that half of the subjects receive the test article under the left axilla and half receive the test article under the right axilla. The quantity of test article application must accurately reflect typical consumer use of that product. Treatment application is made once daily. Test subjects are then placed in a hot room for a 40-min warmup period and two 20-min collection periods. After the warmup, weighed absorbent pads are placed in both axillae for the first 20-min collection period. Immediately following, the pads are weighed again to determine the amount of perspiration collected. The process is repeated for a second 20-min collection period. (The ambient procedure is similar except that the subjects go about their normal daily routines during a 3–5 h collection period.)

A typical antiperspirant protocol consists of the study schedule shown in Table 1. Data treatment consists of performing statistical tests designed to demonstrate whether or not at least 50% of the population obtains a sweat reduction of at least 20% with high probability. (That is the level at which consumer perception of an effect should be noticed.) Antiperspirant activity is evaluated by determining shifts in ratios of the sweat output by the treated axilla to the output of the untreated axilla for each panelist. The data are then analyzed using the Wilcoxon Rank Sum Test, which is recommended in the "Guidelines for Effectiveness Testing of OTC Antiperspirant Drug Products" (5). The source data for this analysis are right-to-left ratios adjusted for 20% reduction due to treatment. These ratios are calculated using the posttreatment average B and C collections for each individual. The adjusted right-to-left ratios for this analysis are calculated as follows: (1) For subjects treated on the right axilla $X = R/(0.8 L)$; and (2) For subjects treated on the left axilla $Y = (0.8 R/L)$.

Table 1 Study Schedule for Typical Antiperspirant Protocol

Day 17 to day 0	Conditioning period
Day 1	Medical screening, baseline sweat collection (optional), test article application no. 1 (test article applied to one axilla, placebo or untreated to the contralateral axilla)
Day 2	Test article application no. 2
Day 3	Test article application no. 3; sweat period 1 h after treatment application no. 3
Day 4	Test article application no. 4
Day 5	Sweat collection 24 h after treatment application no. 4

The hypotheses tested in the rank sum test are stated as follows:

Ho: median $X \geq Y$

Ha: median $X < Y$

Rejection of the null hypothesis will justify the conclusion that at least 50% of the target population will obtain a sweat reduction of at least 20%. An example of the results of a typical antiperspirant study employing this procedure is presented in Fig. 6. Manufacturers whose products meet these guidelines are deemed to have substantiated the claim that their products are antiperspirants.

When baseline measurements are utilized, the guidelines recommend the following analysis utilizing the Wilcoxon Signed Rank Test. The data used for this analysis are the adjusted ratios, Z values, calculated from the pre- and posttreatment average B and C collections for each individual.

SUBSTANTIATION OF PRODUCT DEFINITION										
			1 Hour After Application No. 3				24 Hours After Application No. 4			
			Milligram Collection		Wilcoxon Test		Milligram Collection		Wilcoxon Test	
Subject	Treatment	Axilla Treated	B	C	X	Y	B	C	X	Y
1	T	L	294	335	—	1.63	652	442	—	1.50
	P	R	601	683			1198	859		
2	T	R	184	351	0.69	—	78	175	0.61	—
	P	L	293	671			151	370		
3	T	L	428	553	—	2.23	546	476	—	2.15
	P	R	1200	1545			1512	1240		
⋮										
Wilcoxon Rank Sum Test p-value			<0.01				<0.01			
Estimate of Percent Reduction			44.4%				48.0%			

T = Treatment, P = Placebo

Based on the results of the Wilcoxon Rank SumTest, it can be concluded that at least 50 percent of the target population will obtain a sweat reduction of at least 20 percent. This product meets the guidelines and can be labeled as an antiperspirant.

Figure 6   An example of product definition results for substantiation of an antiperspirant product.

$$Z = (PC \times T) / (PT \times C)$$

Where PC is the pretreatment measure of moisture for the control axilla, PT is the pretreatment measure of moisture for the treated axilla, C is the posttreatment measure of moisture for the control axilla, and T is the posttreatment measure of moisture for the treated axilla. The hypotheses tested in the signed rank test are stated as follows:

Ho: median $Z \geq 0.80$

Ha: median $Z < 0.80$

Rejection of the null hypothesis will also justify the conclusion that at least 50% of the target population will obtain a sweat reduction of at least 20%.

Substantiation of Product Performance Characteristics

Product performance characteristics for antiperspirant products consist of claims such as 48-h protection, lasts longer, etc. For performance characteristic testing, the product definition claim is tested at the performance standard identified. For example, for a claim of 48-h protection, antiperspirant efficacy is measured at 48 h following treatment, as shown in Fig. 7. Study schedule for a typical antiperspirant protocol is shown in Table 2.

Consumer expectation and experience are probably global and not specifically measured at precise increments. However, consumers expect that they will be protected with the implied claim.

Substantiation of Product Attribute Characteristics

These study designs produce claims about specific attributes that may characterize the product more distinctly to the consumer.

For antiperspirant products, attributes such as mildness, ease of application, amount of residue (staining), stinging potential, etc., can be measured by consumers.

A mildness or sting study might consist of having females apply

SUBSTANTIATION OF PRODUCT PERFORMANCE CHARACTERISTICS					
			48 Hours After Product Application			
			Milligram Collection		Wilcoxon Test	
Subject	Treatment	Axilla Treated	B	C	X	Y
1	T	R	204	236	0.73	—
	P	L	348	402		
2	T	R	1123	1341	0.93	—
	P	L	1575	1731		
3	T	L	405	421	—	1.31
	P	R	639	711		
:	:	:	:	:	:	:
		Wilcoxon Rank Sum Test p-value			<0.01	
		Estimate of Percent Reduction			38.6%	

T = Treatment, P = Placebo

Based on the results of the Wilcoxon Rank SumTest, it can be concluded that this product meets the criteria for claiming 48-hour protection.

Figure 7   Product performance at 48 h after product application.

the product immediately after shaving the axillary region. Typically, a test product containing the active ingredient is applied to one axilla and a placebo is applied to the opposite axilla. Test products are randomly assigned and product identification is blinded from the consumer. Table 3 shows a protocol example for an attribute evaluation. A consumer will

Table 2   Study Schedule for a Typical Antiperspirant Protocol

Day 17 to day 0	Conditioning period
Day 1	Medical screening, baseline sweat collection (optional), test article application no. 1 (test article applied to one axilla, placebo or untreated to the other axilla)
Day 2	Test article application no. 2
Day 3	Test article application no. 3
Day 4	Test article application no. 4
Day 5	Nothing
Day 6	Sweat collection 48 h after treatment application no. 4

Table 3 Protocol Example for an Attribute Evaluation

Day 5 to day 0	Prescreening
Day 1	Medical screening. Subject shaves one axilla, applies test product A to the axilla and completes a stinging questionnaire. The other axilla is shaved, test product B is applied, and a second questionnaire is completed.

generally have a predisposed opinion about an attribute and will determine its presence quite easily.

Substantiation of Product Comparative Efficacy Characteristics

Study designs for comparative product efficacy are usually performed to determine where a product ranks in the competitive marketplace. This type study will help a manufacturer determine whether a product is better, worse, or comparable to other brands. The position may influence what other claims are made about the product.

Study designs for antiperspirant products consist of direct comparisons of one product to another using the same subject. The study procedures are the same as in the guidelines except that the subjects will have both test products randomly assigned to their right and left axillae.

A typical protocol design for this type of study is shown in Table 4.

Table 4 Typical Protocol Design for Comparative Study

Day 17 to day 0	Conditioning period
Day 1	Medical screening, baseline sweat collection (optional), test article application no. 1. (Test article A applied to one axilla, test article B applied to the other axilla.)
Day 2	Test article application no. 2
Day 3	Test article application no. 3
Day 4	Test article application no. 4
Day 5	Sweat collection 24 h after treatment application no. 4

Data analysis consists of hypothesis testing to determine whether or not there is a statistically significant difference between the products. Adaptations of the Wooding-Finkelstein (4) analysis method are commonly utilized. In these studies, the number of subjects is dictated by the variability of the data and the expected level of meaningful difference. Figure 8 presents examples of sweat collection data and analysis demonstrating comparative efficacy of two products.

Consumers may independently determine that one product performs differently than another, but this is usually not discovered on a side-by-side basis and certainly not under controlled conditions. Therefore, consumers' in-use perception of a superiority-type claim is one based upon their own set of circumstances.

SUBSTANTIATION OF PRODUCT COMPARATIVE EFFICACY

Subject	Treatment	Axilla Treated	Milligram Collection		Average	Log
			B	C		
1	B	L	1085	632	858.5	6.76
	A	R	925	732	828.5	6.72
2	A	L	416	290	353.0	5.87
	B	R	474	322	398.0	5.99
3	A	L	848	1057	952.5	6.86
	B	R	896	1026	961.0	6.87
⋮	⋮	⋮	⋮	⋮	⋮	⋮

A = Treatment A, B = Treatment B

The analysis of variance method described by Wooding and Finklestein [4] utilizes the log transformation of average B and C collection data. The analysis of variance table is shown below.

	Sums of Squares	Mean Squares	f-value	p-value
Subject	12.33	0.39	29.49	0.0001
Side (Axilla)	0.02	0.02	1.26	0.2711
Test Product	0.08	0.08	6.02	0.0200
Error	0.41	0.01	—	—

Estimate of percent difference between product A and B = 6.67%

Based on the results of the analysis (test product p-value <0.05), it can be concluded that Test Product A has significantly better efficacy than Test Product B.

Figure 8 Sweat collection data and analysis of results comparing products A and B.

Table 5 Typical Protocol Design for Consumer Preference Study

Day 3 to day 0	Conditioning or washout period
Day 1	One-half of the test subjects are given product A to use, one-half are given product B to use
Day 1 to day 5	Products are used as directed
Day 6 to day 8	Test subjects return the product and begin intermediate conditioning period
Day 8	Test subjects are given the other product to use
Day 8 to day 14	Products are used as directed
Day 14	Test subjects return the product they were using and complete a product preference questionnaire

Substantiation of Consumer Preference

These study designs are carried out employing actual consumer users of antiperspirant products.

One test product is given to the consumer to use for a specified time period. After a brief rest period, the consumer crosses over to the other test product for comparison. Following the second use period, the consumer is asked to indicate a preference for one product or the other. A typical protocol for such a study is shown in Table 5.

The advantage of the consumer study is that it provides the sponsor with the consumer's opinion of the selected test product quality or attribute.

DEODORANT STUDY DESIGNS

Substantiation of Product Definition

Deodorant claims for personal care products are considered to be cosmetic claims and are not regulated by any FDA monograph unless antiperspirant and/or antimicrobial claims are also made. So it is left up to the manufacturer to substantiate the product definition claim of deodorancy. The basic study design for deodorant products consists of selecting a panel of subjects with high axillary odor, applying test product, and then measuring the reduction of body odor using a panel of trained odor judges. If there is a statistically significant treatment effect when the product is compared to a placebo or untreated control, then the product has deodorant efficacy.

In 1987 the American Society of Testing and Materials (ASTM) prepared a "Standard Practice for the Sensory Evaluation of Axillary Deodorancy" (6). This ASTM document states that in order for a product to qualify as an effective deodorant, it should meet the criteria established in these guidelines. A summary of these guidelines follows.

Test subjects should represent the consuming population. They should have recognizable body odor at levels set by specific protocols. Subjects with extremely high or low odor may not be acceptable. Subjects with a large difference in odor level between the right and left axillae should be excluded. Subjects participate in a minimum 7-day conditioning period during which they do not use any deodorant products (a longer period may be necessary if they have been using an antiperspirant product). During the test phase, subjects go about their normal daily activities with the following restrictions: no spicy foods, no swimming, no perfumes, and minimal physical activity with no eating or smoking allowed prior to an odor evaluation. These restrictions have been known to affect axillary odor.

Following the conditioning period, a control (baseline) odor evaluation is conducted and subjects are assigned treatment. Treatments are randomly assigned but ensure that half of the subjects receive the active test article under the left axilla and half receive the test article under the right axilla. A placebo test article is applied to the opposite axilla. The quantity of the test article applied must accurately reflect typical consumer use of that product. Treatment applications are made once daily for a number of consecutive days and are followed by posttreatment evaluations usually conducted between 8 and 24 h after the final application. A typical schedule for substantiation of deodorant product definition is shown in Table 6.

Data treatment consists of performing statistical tests designed to determine if there is a significant difference between the test and control products. The posttreatment differences in four-judge average odor scores are analyzed using the distribution free signed rank test. The null hypothesis, which states that the difference between the paired test articles (treatment, control) is equal to zero (Ho: control minus treatment equals zero), will be rejected if the signed rank test p-value is less than or equal to 0.05. For example, data in Fig. 9 exhibit results from a study that would support the definition of deodorancy for a treatment (T) versus a placebo (P) at 12 h following treatment.

Table 6 Typical Schedule for Substantiation of Deodorant
Product Definition

Day 14 to day 1	Conditioning period
Day 1	A.M. Control wash
Day 2	A.M. Control odor evaluation, subject selection, test article application no. 1. (Test article applied to one axilla, placebo or untreated to opposite axilla.)
Day 3	A.M. test article application no. 2
Day 4	A.M. test article application no. 3
Day 4	P.M. 8-h and 12-h odor evaluation

Manufacturers whose products meet these guidelines have substantiated their claim that their products are deodorants.

Substantiation of Product Performance Characteristics

Product performance characteristics for deodorant products involve claims such as all-day protection, 4-, 8-, 12-, or 24-h protection, lasts longer, etc.

SUBSTANTIATION OF PRODUCT DEFINITION									
	12 HOURS AFTER PRODUCT APPLICATION NO. 3								
	Judge 1		Judge 2		Judge 3		Judge 4		4-Judge Average	
Subject	T	P	T	P	T	P	T	P	T	P
1	1	3	2	4	1	4	3	6	1.8	4.3
2	3	7	2	6	2	4	3	5	2.5	5.5
3	1	3	1	3	1	3	1	2	1.0	2.8
⋮	⋮	⋮	⋮	⋮	⋮	⋮	⋮	⋮	⋮	⋮
						Mean		1.6	4.0
					Mean difference			2.42	
				Signed rank test p-value				<0.05	

T = Treatment, P = Placebo

The results of the statistical analysis on the 4-judge average differences between treatment and placebo indicate a statistically significant treatment effect. Based on these results the test product can be labeled as having deodorant efficacy.

Figure 9 Data on deodorant efficacy.

For performance characteristics testing, the product definition claim is designed to identify the selected performance characteristic. For example, for a claim of 24-h protection (Fig. 10), deodorant efficacy is measured at 24 h following treatment using the same protocol design as above.

Consumer expectation and experience are usually more global and not specifically measured at precise intervals. However, consumers expect that they will be protected with the implied claim.

Substantiation of Product Attribute Characteristics

These study designs allow the manufacturer to make a specific sensory attribute claim. For deodorant products, consumer response to any self-perceived characteristic—such as ease of application, fragrance, creaminess, stickiness, and overall liking—can be measured using a variety of scales.

This information is collected by administering a self-perception questionnaire following a specified use period. Figure 11 is an example of such a questionnaire.

For consumers, studies of attribute characteristics are more easily

SUBSTANTIATION OF PRODUCT PERFORMANCE CHARACTERISTICS (i.e. 24-Hour Protection)									
	24 HOURS AFTER PRODUCT APPLICATION NO. 4								
	Judge 1		Judge 2		Judge 3		Judge 4		4-Judge Average	
Subject	T	P	T	P	T	P	T	P	T	P
1	3	6	4	6	4	7	3	6	3.5	6.3
2	4	5	2	5	2	6	2	6	2.5	5.5
3	5	6	5	7	6	6	5	5	5.3	6.0
⋮	⋮	⋮	⋮	⋮	⋮	⋮	⋮	⋮	⋮	⋮
							Mean		4.2	5.9
						Mean difference			1.72	
						Signed rank test p-value			<0.05	

T = Treatment, P = Placebo

The results of the statistical analysis on the 4-judge average differences between treatment and placebo indicate a statistically significant treatment effect. Based on these results the test product can be labeled as having 24-hour deodorant efficacy.

Figure 10 Data showing product performance at 24 hours following treatment.

Self-Perception Questionnaire   Name: _____

WHEN ANSWERING THE FOLLOWING QUESTIONS PLEASE CONSIDER YOUR **OVERALL** IMPRESSION OF THE TEST PRODUCT.

1.)   How would you describe the **ease of application** of the product?

    ( ) Extremely easy
    ( ) Very easy
    ( ) Moderately easy
    ( ) Slightly easy
    ( ) Not at all easy

2.)   How would you describe the **fragrance** of the product?

    ( ) Too strong
    ( ) Slightly too strong
    ( ) Just right
    ( ) Slightly too weak
    ( ) Too weak

3.)   How would you describe the **creaminess** of the product?

    ( ) Extremely creamy
    ( ) Very creamy
    ( ) Moderately creamy
    ( ) Slightly creamy
    ( ) Not at all creamy

4.)   How would you describe the **stickiness** of the product?

    ( ) Extremely sticky
    ( ) Very sticky
    ( ) Moderately sticky
    ( ) Slightly sticky
    ( ) Not at all sticky

5.)   How much did you like or dislike the product **overall**?

    ( ) Like extremely
    ( ) Like moderately
    ( ) Neither like or dislike
    ( ) Dislike moderately
    ( ) Dislike extremely

Figure 11   Product attribute self-perception questionnaire.

related to their actual experience. They usually have a predisposed attitude toward an attribute and will determine the presence of these attributes quite readily.

Substantiation of Product Comparative Efficacy Characteristics

Studies for product comparative efficacy are usually designed to determine where a product ranks in the competitive marketplace. That position may influence what claims are made about the product.

SUBSTANTIATION OF PRODUCT COMPARATIVE EFFICACY									
	8 HOURS AFTER PRODUCT APPLICATION NO. 4								
	Judge 1		Judge 2		Judge 3		Judge 4		4-Judge Average	
Subject	T-1	T-2	T-1	T-2	T-1	T-2	T-1	T-2	T-1	T-2
1	0	2	1	2	0	2	1	2	0.5	2.0
2	2	2	2	3	3	2	1	2	2.0	2.3
3	1	2	2	2	1	1	1	1	1.3	1.5
⋮	⋮	⋮	⋮	⋮	⋮	⋮	⋮	⋮	⋮	⋮

	T-1	T-2
Mean	1.3	1.9
Mean difference	0.52	
Signed rank test p-value	<0.05	

T-1 = Test Product One, T-2 = Test Product Two

The results of the statistical analysis on the 4-judge average differences indicates a statistically significant difference between the two treatments. Based on these results the manufacturer of Test Product One can claim superior efficacy over Test Product Two.

Figure 12   Data showing comparative efficacy of two deodorant products 8 h following treatment.

Study designs for deodorant products consist of direct comparisons of one product to another using the same subject. Figure 12 presents data comparing two treatments. The study procedures are the same as in the guidelines except that the subjects will have both test products randomly assigned to the right and left axillae. Protocol design and data analysis are the same as for substantiation of product definition (treatment versus control). In these studies, the number of subjects is dictated by the variability of the data and the expected level of meaningful difference.

Substantiation of Consumer Preference

These studies are conducted with actual consumer users of deodorant products. One test product is given to the consumer to use for a specified time period. After a brief rest period, the consumer crosses over to the other test product for comparison purposes. Following the second use period, the consumer is asked to indicate a preference for one product for the identified attributes. The protocol design for substantiation of consumer preference for deodorant products would be the same as that

Preference Questionnaire   Name: _____

WHEN ANSWERING THE FOLLOWING QUESTIONS PLEASE COMPARE YOUR **OVERALL**
IMPRESSION OF THE FIRST TEST PRODUCT TO THE ONE YOU USED SECOND.

1.)   Which product did you prefer for the **Ease of Application?**

    ( ) Product used first
    ( ) Product used first

2.)   Which product did you prefer for **Fragrance?**

    ( ) Product used first
    ( ) Product used first

3.)   Which product did you prefer **Overall?**

    ( ) Product used first
    ( ) Product used first

Figure 13   Product attribute preference questionnaire.

for antiperspirant products. Figure 13 is an example of a preference questionnaire.

PARITY CLAIMS

We could not write a chapter on claims substantiation without mentioning parity claims. There are times when manufacturers want to state that their product is "as good as" the competitor's or that theirs is "unsurpassed." Protocol designs for these studies are similar to those presented previously for comparative efficacy; however, the study design requires more detail prior to initiation. To claim superiority, a manufacturer must show a statistically significant difference between his product and the competitor's. A parity position is substantiated by the failure to find such a difference. However, the power of the test must be sufficiently great to find a meaningful difference if one exists. Conducting a study with a small number of test subjects and finding no statistically significant difference is not sufficient for a parity claim. The study design must be able to withstand any criticism that the investigator did not "look hard enough" to find a difference or that the sample size and power calculations cannot support this premise. We recommend that you review "Substantiating a Parity Position" (7), as it outlines the necessary preparation.

FORUMS FOR REVIEW, CHALLENGE, AND RECOURSE

A challenge can occur in any of the areas mentioned earlier. Radio and television network review takes place prior to the airing of a commercial and usually involves clarification of the commercial's story with the supportive data being submitted to the network for their approval prior to airing. Study investigators and other experts have been called upon to comment on the meaning or implication of a study primarily as it relates to the consumer's potential perception of the claimed benefit or attribute (8).

Challenges that invoke the Lanham Trademark Act are usually the most serious forms of challenge. These challenges are directed by court actions being brought by one manufacturer against another directly relating to a claim being made about a product. The implications of these cases are very significant and can result in millions of dollars in damages and or costs depending on the court action. The outcomes of these cases may range from changes in the advertising and/or packaging to total withdrawal of the product from the market and potential punitive damages. Testing support has had a tremendous effect on the final outcome of Lanham Act cases.

The most common route for challenge of an advertising claim is the National Advertising Division of The Council of Better Business Bureaus, Inc. (NAD). Its stated purpose is as follows: "NAD and the Children's Advertising Review Unit (CARU), in association with the National Advertising Review Board (NARB), seek to sustain truth and accuracy in national advertising through a self-regulatory program designed by the advertising community and administered by the Council of Better Business Bureaus, Inc." NAD Case Reports can be used as references to determine how NAD reaches their decisions and what and how substantiation data was used to achieve each claim (9).

CONCLUSION

The antiperspirant and deodorant marketplace is a very competitive arena. Manufacturers are eager to design claims that protect market share or enlarge it. Changing technology allows for the formulation of various new product forms that better fit the changing lifestyles of consumers. While products may appear to be more esthetically pleasing with perceptions of purity and additional long-lasting protection, the emerging sensitive-skin and residue-averse markets also affect the claims to be made. In designing tests to support such claims, marketers, research and development, and advertisers must be communicating in order to be prepared for all these factors (10).

REFERENCES

1. Laden K, Felger C. Antiperspirants and Deodorants. New York: Marcel Dekker, 1988, Chap 1.
2. Jungermann E. Clear antiperspirant stick technology: A review. Cosmet Toiletr 1995; 110:49–56.
3. Department of Health and Human Services, Food and Drug Administration. Antiperspirant drug products for over-the counter human use: Tentative final monograph. Fed Reg 1982; 47(162):36491–36505.
4. Wooding WM, Finkelstein P. A critical comparison of two procedures for antiperspirant evaluation. J S Cosmet Chem 1975; 26:225–275.
5. Department of Health and Human Services, Food and Drug Administration. Guidelines for effectiveness testing of OTC antiperspirant drug products. Fed Reg 1982; 47.
6. ASTM Committee E-18. Standard Practice for the Sensory Evaluation of Axillary Deodorancy, ASTM designation: E 1207, 1987; 15.07.
7. Buchanan B, Smithies R, Substantiating a parity position. J Adv Res October/November 1989.
8. Department of Health, Education and Welfare, Food and Drug Administration. Obligations of clinical investigators of regulated articles: Proposed rule. Fed Reg 1978; 43(153):35209–35236.
9. Council of Better Business Bureaus, National Advertising Division. NAD Case Reports 1996; 25:253.
10. Jungermann E. Antiperspirants: New trends in formulation and testing technology. J Soc Cosmet Chem 1974; 25:621–638.

8

Mildness Testing for Personal Washing Products

Richard I. Murahata and Gregg A. Nicoll
Unilever Research U.S., Edgewater, New Jersey

INTRODUCTION

Mildness benefit has become an increasingly important component of claims for personal washing products. Personal washing products, by definition, must first deliver a cleansing benefit; however, the trend in formulation directly reflecting an expressed consumer desire has been toward cleansers that are mild (1). While all commercially available washing products are safe to use and none are "harsh" in the absolute sense, there are differences among products that can be measured both by clinical/instrumental assessment and consumer perception. Many companies treat the exact test protocols as proprietary, but there are some that have been published in the peer-reviewed scientific literature and other "generic" methods, which have been explicitly or implicitly indicated as appropriate. This chapter concentrates on test methodologies that have been used or could reasonably be used in support of mildness testing. While some claims are based on a single source of data, many have taken the "three-pronged" approach, comprising expert clinical evaluation, self-assessment, and instrumental measurements. While the first two may stand alone, the latter rarely does except in specific circumstances, discussed under "Instrumental Techniques," below. This review is not intended to be comprehensive but rather to present references illustrative of the particular points.

There is a general relationship between the aggressiveness of a test method and the ability to discriminate among products, as shown in Fig. 1. Methodologies that are extremely mild (i.e., "normal use") tend to distinguish those materials that are relatively harsh from the remainder. This is especially true when expert evaluation is used as the sole criterion of judgment. The use of self-assessment (e.g., subjective responses of itching and tightness) as well as sensitive instrumental assessments (e.g., barrier damage) can often demonstrate differences even under these extremely mild conditions. The discrimination can also be enhanced, often at great expense, by increasing the sample size to hundreds or even thousands of subjects. Discrimination among products may also be increased by moving along the aggressiveness axis. In the middle are mildly exaggerated procedures (i.e., repeated arm wash or face washing, multiple hand soaks, and the more aggressive flex wash). These methods are able to detect differences in products over a relatively wide range of mildness. At the other extreme, the soap chamber

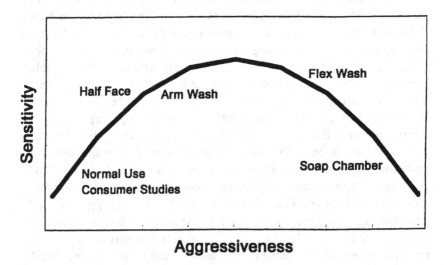

Figure 1 Depiction of the observed relationship between aggressiveness of test procedure and ability to discriminate differences.

irritation test (2), a relatively aggressive procedure, tends to distinguish the mild products from the rest of the cleansers. Rarely have different test protocols been directly compared, and it has been a challenge to link the relationship among the various published test methodologies. Nicoll et al. (3) directly compared the relative sensitivity of two arm-wash methods, while others have shown a reasonable correlation to exist between results of flex wash and consumer perception (4,5) or an arm wash and consumer perception (7).

IN VIVO METHODS

Various in vivo use tests have been developed for evaluating the relative mildness of personal cleansing products on human skin, including patch testing, immersion-type tests, mild-to-moderate exaggerated washing tests, and normal-use tests (2,6–17). Exaggerated washing tests were developed to emulate, to varying degrees, normal use of personal cleansing products; several have been shown to be good indicators of product mildness and the consumer perceptions that develop under normal in-home use conditions (6,15). Exaggerated washing tests typically are of shorter duration and involve fewer subjects compared to normal-use studies; this allows for the evaluation of a greater number of formulations while maintaining sensitivity to discriminate between products of similar mildness potential.

Each of the general types of clinical test methods described have different advantages for characterization of the effects of personal cleansing products on skin (Tables 1 and 2).

Exaggerated-Use Tests

Four general types of mildly exaggerated washing tests have been described for use in the evaluation of personal cleansing products: (1) arm-wash tests, which involve multiple daily washes of the volar aspect of the forearm for 5 consecutive days (6,13–15); (2) half-face tests, wherein panelists wash two to four times daily during a 5-day study period (16); (3) flex-wash tests, which entail washing the flex area (antecubital fossa) of the arm with a sponge three times daily for 5 days (12);

Table 1 Benefits and Limitations of the Various Types of Clinical Test
Protocols Used to Evaluate Personal Cleansing Products

Test	Benefits	Limitations
Patch tests	Multiple product intra-subject comparisons. May predict longer-term use studies. Involves low numbers of subjects (<30), less than 1 week in duration, and requires small amounts of sample.	Greatly exaggerated exposure conditions. Difficult to account for rinsing.
Exaggerated-use tests	Predictive of normal-use tests. Involves low numbers of subjects ($n = 20$–40) and is short in duration (<3 weeks).	Uses exaggerated-use wash conditions.
Normal-use tests	Test conditions that emulate the situation test products under which anticipated to be used.	Requires large numbers of subjects using test products over extended periods of time.

and (4) hand-wash tests, wherein subjects perform multiple hand wash-ings daily typically for a week (7). Another method that has been used for evaluating personal cleansing products is the leg-wash test, wherein subjects wash the lateral aspect of the lower legs. This test protocol is similar to the half-face test except for the test site. There are no refer-ences to this method in the literature.

There are several advantages to these types of tests. Single mate-rials, simple surfactant systems, and fully formulated products of vari-ous forms including bars, liquids, and foams can be tested with little or no modification of the basic protocol. The test methods have been designed to incorporate "wash" procedures that emulate, to varying

Table 2 Assessment of the Effects of Cleansing Products on Skin by Exaggerated-Use–Test Methods

Test	Claim	Skin condition assessed	Expert assessment	Instrumental assessment	Panelist self-assessment
Flex-wash test	Mildness	Irritation	Yes	Limited	No
Arm-wash test	Mildness; dryness potential	Irritation, dryness	Yes	Yes	Limited
Half-face test	Mildness; dryness potential	Irritation, dryness	Yes	Yes	Yes
Hand-wash test	Mildness; dryness potential	Irritation, dryness	Yes	Limited	Yes

degrees, consumer habits. Factors such as rinsability, solubility, and inclusion of skin benefit agent (e.g., glycerol) delivery, can be accounted for by use of the appropriate conditions. Since many of these tests utilize a paired-comparison study design, each subject serves as his or her own control. These methodologies are currently in widespread use within the industry for supporting mildness claims.

Flex Wash

Of the published methods, the flex-wash (11) test appears to exhibit the greatest capability to distinguish between products based on differences in skin irritation potential. Briefly, this test consists of three daily 1- or 2-min washes of the antecubital fossa (flex area) of the arm. Studies testing products believed to be relatively harsh utilize a 1-min wash cycle to protect the subjects from unnecessary discomfort, whereas flex washes used to evaluate mild products use a 2-min wash protocol. Washing is conducted for a maximum of 5 consecutive days. The flex-wash test is not used to evaluate products based on skin drying potential, since the sponge mechanically removes flaking skin, a criterion used to evaluate the skin-drying potential. Since dryness is not a clinical parameter being assessed, the flex wash is less sensitive to climatic conditions than an arm wash or half-face test. Consequently, the flex wash can be conducted year-round.

A key limitation of the flex wash test results from the use of a wash implement—i.e., a sponge. As mentioned above, the sponge removes the visible signs of skin dryness, the uplifting skin flakes or scales. And secondly, the sponge can cause irritation to the skin when used in an inappropriate manner (6). For example, use of a dry sponge directly on the skin or testing a material with little lubricity (e.g., with short-chain alcohols or water) results in the generation of misleading data. Self-assessment of skin condition by the flex-wash test panelists has not been demonstrated to provide evaluation advantages beyond those obtained by expert assessment alone.

Arm-Wash Tests

Arm-wash tests were developed to more closely emulate normal consumer habits and have been shown to be a good indicator of the product mildness attributes experienced under normal in-home use conditions (6,17). Arm-wash tests can distinguish products based on the differences in both skin drying and irritation potential. However, discrimination based on irritation potential alone may be reduced compared to a flex-wash test. Importantly, the results of arm-wash tests are directionally similar to those obtained using flex-wash and half-face tests. The use of the volar forearm as the anatomic test site more readily permits the use of instrumental techniques to assess changes (e.g., barrier damage) in the skin as well as simple tape stripping to obtain samples for biochemical analysis. Arm-wash tests, therefore, not only allow comparison of the irritation potential and skin drying potential of skin cleansing products but also more easily permit the investigation of the effects on skin function, physiology, and biochemistry.

The skin dryness component of the arm-wash test is especially sensitive to ambient weather conditions. These tests are typically conducted when the average daily temperature is less than 50°F and the average daily relative humidity less than 50%. Studies have been conducted where the temperature and humidity are higher than 50°F/50% RH, but extreme care must be taken in interpreting the results, especially if the test products being compared are of similar mildness potential.

An arm-wash test typically consists of washing of the volar surface of the forearm four times daily for 4 consecutive days and two washes on the fifth day. Different wash-cycle durations have been reported ranging from a 10-s wash/90-s leave-on to a 2-min wash cycle (6,13). As discussed further on, the duration of the wash cycle significantly affects the sensitivity of the test method. Inclusion of instrumental measurements of skin condition with clinical assessment procedures may further enhance the power of discrimination and may provide insight into the functional changes in skin condition occurring during the study (13).

Self-assessment by the panelists of their own skin condition can be included in the arm-wash test and provides information supplementing expert assessment of dryness and irritation. Self-assessment permits the evaluation of skin responses not measurable by expert assessment (e.g., itching, burning, stinging). Self-assessment may add less value in an arm-wash test compared to that gained by its use in a half-face test.

Comparison of Two Different Arm-Wash Methods

It may be of interest to compare how the washing procedure can influence the sensitivity of the test methodology (3). Two different arm-wash test methods have recently been reported (6,13). Both consist of exposing the volar surface of the forearm to a product 18 times during a 5-day test period (four times per day for 4 days and twice on the fifth day). The primary difference between the two protocols is in the method and duration of product application. Sharko et al. (1) used a gentle washing action for 2 min prior to rinsing. In contrast, Keswick et al. (6) used a 10-s product application followed by a 90-s exposure to the lather before rinsing. As expected, increased washing times associated with the first method resulted in greater absolute levels of erythema, dryness, and instrumentally assessed damage compared to the second method. More importantly, the relative sensitivity of the test method is also influenced by the wash procedure, with the first method providing greater power of discrimination.

Half-Face Test

A half-face test can distinguish products based on the differences in both skin drying and irritation potential. However, discrimination based on clinical assessment of irritation potential may be reduced compared to the flex-wash test. A key advantage of the half-face test is that this method readily permits the self-reporting of sensory (e.g., burning, stinging, after-wash tightness) changes in skin condition that are less easily detected in an arm-wash test and poorly detected by use of the flex-wash test.

A half-face test consists of a wash of the cheek area with a gloved hand or nonwoven cotton pad four times daily for 5 days (Murahata et al., unpublished data, and Ref. 17). A single product is assigned to each side of the face and a wash technician gently washes both right and left sides of the face simultaneously. Expert clinical assessment of erythema, dryness/scaling, and roughness are commonly done, as is self-assessment. Instrumental measurements may be taken, but it is technically more difficult to obtain accurate readings from the face compared to the forearm.

Key limitations of a half-face test are weather (similar to arm-wash methodologies), facial hair, and the requirement for the study subjects not to use any cosmetics or moisturizing products during the study.

Correlation of Methods

The flex-wash, arm-wash and half-face test methods have provided product comparison results that were consistent with each other (Nicoll et al., unpublished data). The flex-wash test is predictive of the erythema that could develop when the same products are evaluated by the arm-wash and half-face tests. This correlation for irritancy potential among the flex-wash test, an arm-wash test, and a half-face test is evident when either bar or liquid forms of personal cleansing products are evaluated. This correlation among the three methods is also evident when products that employ a soap or a synthetic surfactant (also known as a synthetic detergent or syndet) as the primary active system are compared. These results are consistent with other studies that report the usefulness of exaggerated washing studies for comparing a soap to a syndet bar, and a syndet bar to a syndet liquid (6,17).

A hand-wash test (8) has been partially compared with the other exaggerated-use tests described above, and a full comparison would be anticipated to generate data consistent with the other exaggerated-use test methods.

Each of the clinical test methods discussed here provides distinct benefits for characterization of the effects of personal cleansing products on skin, and, when used in combination, have the potential to provide a powerful package for substantiated mildness claims.

SENSORY SIGNALS AND SKIN MILDNESS

We have seen that a number of test methodologies have been used to demonstrate the skin interaction properties of mild cleansing products. In many of these tests, skin condition is monitored subsequent to the subject's experiencing a period of exaggerated use. Exaggerated use is generally required to elicit a significant and observable response from the skin so that differences in response between products can be either scored visually or quantified instrumentally. While testing of this type is widely recognized for its contribution to the support of advertising and patent claims, the relevance of these tests to consumer perception has not always been appreciated. A recent study has tested a wide variety of cleansing bars by several different test methods (4,5).

The first test, mentioned by Celleno et al., is a modified Draize patch test in which skin irritation is monitored following a 48-h patch of 2% solutions of various cleansing bars. Celleno et al., found that none of the products tested showed a significant response to this degree of insult. When this test was repeated using the rather more rigorous soap-chamber test recommended by Frosch and Kligman (2), all products produced a significant skin response. It was then possible to rank the products tested in terms of irritancy and to divide the products into several groups. Celleno et al. also showed that the ranking obtained from the soap-chamber test agreed with the ranking yielded by a flex-wash test.

In addition to a series of exaggerated tests, Celleno et al. also performed a test in which subjects were asked to wash their faces at home, twice a day and in a normal fashion (5). Subjects kept a log in which they recorded the incidence of either positive signals (hydration, softness, smoothness) or negative signals (dryness, tension, roughness). By this means, the frequency of various sensory signals following a normal wash could be compared to the irritation potential predicted by an exaggerated test. The results of this test made it clear that those products that were ranked with a relatively low irritation potential in the exaggerated tests were also more likely to elicit positive sensory signals and unlikely to elicit negative sensory signals.

One of the more popular claims today is that a product is hypoallergenic. There is as yet no industry or regulatory standard for supporting this claim. One manufacturer has used a human repeat insult patch test (HRIPT) with approximately 200 subjects without any positive reactions to support a claim of "produces no evidence of allergic reaction or sensitization" (18).

INSTRUMENTAL TECHNIQUES

The results obtained from using various biomedical instruments have been used, and occasionally misused, in support of specific claims for personal washing products. The need for instrumental assessment has been driven by a desire for increased "objectivity," reproducibility, and sensitivity, as well as the desire to relate biophysical measurements to biochemical and biological measurements. The instruments have been most valuable in those areas where the measurement techniques are robust and well understood, interpretation of the data is relatively unambiguous, and the information adds sensitivity to clinical or self assessment or provides information not obtained by other means. Care must be taken, as there may be a tendency for the data to be overinterpreted, especially when a single measurement is being used to infer a relationship with a complex biological process or where the measurement technique itself has inherent experimental uncertainties. In addition, most instruments in general use today sample relatively small areas. Multiple sampling and/or careful selection of representative areas may be necessary. A further difficulty has been the fact that the optimal value for many of these measurements is not simply the highest or the lowest but some value in between that also needs to be experimentally verified. For example, increasing blood flow may be a sign of irritation, but reducing it to zero is obviously not desirable either.

Since one of the principal roles of the skin is to modulate flow of materials in and out of the body, alteration of normal barrier function has often been used as an indicator of product mildness. Transepidermal water loss (TEWL) has been used successfully by many groups to document the effects on barrier function of the skin. Control of environmental variables is crucial to reproducible measurements, and the key parameters and recommended controls have been published (20). This technique is possibly best suited to measuring barrier damage, such as irritant contact dermatitis, as the result of repeated exposure to a mild irritant (21–28). The sensitivity of TEWL as a screening technique for early signs of irritancy has been reported to be superior to both visual scoring and various biophysical methods (21,27,29–34). As might be expected, increase in TEWL did not correlate with visual scoring of irritation due to application of materials (i.e., DMSO) that are presumed to act below the barrier (35). High baseline (i.e., pre-exposure) values of TEWL were found to be predictive of susceptibility to irritation by some investigators but not by others (24–26,29). Differing test methodologies make direct comparison of the results impossible, and this difference has yet to be resolved.

Direct measurement of skin hydration has been the "holy grail" of biomedical researchers working in this area. Technical advances have built on the basic systems reviewed by Potts (36) and Salter (37), including electrical parameters, mechanical deformation, photoacoustic spectroscopy, and FTIR. Near infrared measurements have been used to examine bound and free water and their relationship to clinical dry skin (38). While good correlative data with visual assessment have been obtained and can play a strong supporting role, there is still sufficient uncertainty in the actual measurement. Care must be taken in using this type of data as the sole evidence for changes in skin moisture. The complexity of the information derived from this class of measurement techniques was nicely described by Salter (39), who opined that "working the microwave region (GHz) may be capable of giving information if hydration is the sole concern" (40,41). Recent work conducted by Salter (42) and Querleux (43) represents approaches of more direct means of quantifying water through magnetic resonance imaging; however, there are severe technical limitations to the spatial resolution of the technique, and it does not appear likely that the water content of the stratum corneum will be directly quantified in vivo in the near future.

Several methods of measuring the remitted light have been used to quantify skin color, especially with respect to irritant-induced erythema (44–46). Perhaps the most commonly used instrument is the Chroma Meter from Minolta. Of the various color systems used for measurement, the L*a*b* system recommended by the CIE (Commission Internationale de l'Eclairage) (47) has found relatively widespread acceptance. The individual values represent a location in a three-dimensional color space. The dimensions are described by L*, representing luminance or brightness, with black being L* = 0 and white being L* = 100; a*, representing the balance between red and green, with a* = 100 being red only and a* = −100 being green only; and b*, representing the balance between blue and yellow with b* = 100 being blue only and b* = −100 being yellow only. As the L*a*b* scale was devised to mimic human perceptual space, it is not surprising that no great increase in sensitivity has been reported. However, the data are quite useful as a confirmatory basis to established quantitative standards and to provide parametric data that can easily be analyzed. With the rapid advances in digital image capture and processing, it can be anticipated that color analysis of large areas of the body will be routine in the near future.

Erythema due to irritation has also been indirectly quantified using laser Doppler velocimetry to measure changes in blood flow (48). Recent advances permit measurement of areas as large as an entire arm or face by using a scanning instrument (49).

Approaches for quantifying skin surface appearance, either noninvasively or using minimally invasive techniques, have been quite successful (for a recent review, see Ref. 50). Image analysis of silicone replicas has been used successfully to quantify changes in skin topography (generally fine lines and wrinkles) either as a result of insult by surfactants or effacement by topical treatments (retinoic acid). Determination of more subtle features (i.e., overall texture) has been much less amenable to routine treatment, and there is probably no generally accepted method for obtaining this type of information. Advances in image capture and novel application of emerging analytical techniques (i.e., fractal analysis) may provide future advances.

Mildness and Rinsability

The connection between mildness and rinsability must be approached with caution, as issues exist with some test methodologies and there has been no clearly demonstrated association between these two parameters. Several authors have proposed the use of fluorescein deposition from aqueous bar slurries spiked with this probe as a valid method for quantifying soap (surfactant) deposition (51,52). However, the validity and relevance of this test methodology has been called into question (53). Fluorescein does not track anionic surfactants, and its intrinsic interaction with skin is highly pH-dependent. In addition to site and amount of binding, the nature of the material must be taken into consideration. Certainly for a given surfactant a good correlation would be predicted between mildness and residue molecularly bound to proteins in the skin. However, modern personal washing products are often formulated with ingredients designed to leave a deposit and provide benefit. Such materials deposited in small amounts may provide a benefit to the appearance and texture of the skin. Other authors have attempted to visualize the deposition through transfer of dye-binding material (54) and related it to an indirect mildness benefit (i.e., reduction of residue transferred to contact lenses).

Elastic properties have been used to characterize dry facial skin (55); a variety of other biophysical techniques have also been used (56).

Whenever possible, a combination of techniques relying on different measurement modalities is recommended. This can be expert clinical assessment, instrumental measurements, and self-assessment, all conducted within a single well-designed study. This design has the added benefit of allowing for the future refinement of the first two techniques to more accurately reflect consumer-perceivable responses.

As we have seen in this limited review, there are a variety of tests available to assess the mildness of personal washing products. The more robust methodologies are often well-controlled laboratory or clinical studies utilizing multiple assessment techniques. This information has been used to guide product development, support claims to medical professionals, and provide the basis for consumer advertising. The success of these techniques will ultimately be tested in the marketplace, where consumers will vote with their dollars and validate the predictive power within the constraints of advertising, brand positioning, and other market drivers.

REFERENCES

1. Murahata RI, Sharko PT, Greene AP, Aronson MP. Cleansing bars for face and body—In search of mildness. In: Rieger M, Rhein LD, eds. Surfactants in Cosmetics. New York: Marcel Dekker, 1997:307–330.

2. Frosch PJ, Kligman AM. The soap chamber test—A new method for assessing the irritancy of soaps. J Am Acad Dermatol 1979; 1:35–41.

3. Nicoll GA, Murahata RI, Grove GL, et al. The relative sensitivity of two arm wash test methods for evaluating the mildness of personal washing products. J Soc Cosmet Chem 1995; 46:129–140.

4. Celleno L, Vaselli A, Ciarrocchi C, et al. Valutazione dermatologica dei prodotti per la detersione della cute. Cosmet Dermatol 1989; 29:19–75.

5. Celleno L, Mastroianni A, Vasselli A, et al. Dermatological evaluation of cosmetic products for skin detergency. J Appl Cosmet 1993; 11:1–22.

6. Keswick BH, Ertel KD, Visscher MO. Comparison of exaggerated and normal use techniques for assessing the mildness of personal cleansers. J Soc Cosmet Chem 1992; 43:187–193.

7. Frosch PJ. Irritancy of soaps and detergent bars. In: Frost PH, Horwitz SN, eds. Principles of Cosmetics for Dermatologists. St. Louis: Mosby, 1982:5–12.

8. Mills OH Jr, Swinyer LJ, Kligman AM. Assessment of irritancy of cleansers. Scientific exhibit presented at the Annual Meeting of the American Academy of Dermatology, Washington, DC, 1984.

9. Sauermann G, Doerscherner U, Hoppe U, Wittern P. Comparative study of skin care efficacy and in-use properties of soap and surfactant bars. J Soc Cosmet Chem 1986; 37:309–327.

10. Komp B. Skin Compatibility tests—Importance in skin cleansing product development. Cosmet Toiletr 1987; 102:89–94.

11. Komp B, Reng, AK. Developing ether sulfate-free surfactant formulations. Cosmet Toiletr 1989; 104:41–45.

12. Strube DS, Koontz SW, Murahata RI, Theiler RF. The flex wash test: A method for evaluating the mildness of personal washing products. J Soc Cosmet Chem 1989; 40:297–306.

13. Sharko PT, Murahata RI, Leyden JJ, Grove GL. Arm wash with instrumental evaluation—A sensitive technique for differentiating the irritation potential of personal washing products. J Dermatol Clin Eval Soc 1991; 2:19–26.

14. Lukacovic MF, Dunlap FE, Michaels SE, et al. Forearm wash test to evaluate the clinical mildness of cleansing products. J Soc Cosmet Chem 1988; 39:355–366.

15. Dahlgren RM, Lukacovic MF, Michaels SE, Visscher MO. Effects of bar soap constituents on product mildness. In: Baldwin A, ed. Proceedings of Second World Conference on Detergents: Looking toward the 90's. New York: American Oil Chemists Society, 1987:127–134.

16. Keswick BN, Visscher MO, Levine MJ. Methods used to assess skin

response to personal cleansers. Scientific exhibit presented at the Annual Meeting of the American Academy of Dermatology, 1991.

17. Doughty D, Jaramillo J, Spengler E. Methods for assessing the mildness of facial cleansing products. In: Preprint of the 16th IFSCC International Congress. 1990:468–477.

18. Clinical Support for Oil of Olay Sensitive Skin Bath Bar. The Procter and Gamble Company, 1994.

19. Kawai M, Imokawa G. The induction of skin tightness by surfactants. J Soc Cosmet Chem 1984; 35:147–156.

20. Pinnagoda J, Tupker RA, Agner T, Serup J. Guidelines for transepidermal water (TEWL) measurements. Contact Derm 1990; 22:164.

21. Tupker RA, Pinnagoda J, Coenraads PJ, Nater JP. The influence of repeated exposure to surfactants on the human skin as determined by transepidermal water loss and visual scoring. Contact Derm 1989; 20:108–114.

22. Tupker, RA, Coenraads PJ, Pinnagoda J, Nater JP. Baseline transepidermal water loss (TEWL) as a prediction of susceptibility to sodium lauryl sulphate. Contact Derm 1989; 20:265–269.

23. Pinnagoda J, Tupker RA, Coenraads PJ, Nater J. Prediction of susceptibility to an irritant response by transepidermal water loss. Contact Derm 1989; 20:341–346.

24. Murahata RI, Crowe DM, Roheim JR. The use of transepidermal water loss to measure and predict the irritation response to surfactants. Int J Cosmet Sci 1986; 8:225–231.

25. Pinnagoda J, Tupker RA, Coenraads PJ, Nater JP. Prediction of susceptibility to an irritant response by transepidermal water loss. Contact Derm 1989; 20:341–346.

26. Agner T. Basal transepidermal water loss, skin thickness, skin blood flow and skin colour in relation to sodium-lauryl-sulphate-induced irritation in normal skin. Contact Derm 1991; 25:108–114.

27. Sharko PT, Murahata RI, Leyden JJ, Grove GL. Arm wash with instrumental evaluation—A sensitive technique for differentiating the irritation potential of personal washing products. J Derm Clin Eval Soc 1991; 2:19–28.

28. Finkey MB, Crowe DM. The use of evaporimetry to evaluate soap induced irritation of the face. Bioeng Skin 1988; 4:311–321.

29. Freeman S, Maibach HI. Study of irritant contact dermatitis produced by repeat patch test with sodium lauryl sulphate and assessed by visual methods, transepidermal water loss and laser Doppler velocimetry. J Am Acad Dermatol 1988; 19:496–502.

30. Wilhelm K-P, Surber C, Maibach HI. Quantification of sodium lauryl sulfate irritant dermatitis in man: Comparison of four techniques: Skin color reflectance, transepidermal water loss, laser Doppler flow measurement and visual scores. Arch Dermatol Res 1989; 281:293.

31. Agner T, Serup J. Sodium lauryl sulphate for irritant patch testing—A dose-response study using bioengineering methods for determination of skin irritation. J Invest Dermatol 1990; 95:543–547.

32. Klein G, Grubauer G, Fritsch P. The influence of daily dish-washing with synthetic detergent on human skin. Br J Dermatol 1992; 127:131–137.

33. Tupker RA, Pinnagoda J, Coenraads PJ, et al. Evaluation of hand cleansers: Assessment of composition, skin compatibility by transepidermal water loss measurements, and cleansing power. J Soc Cosmet Chem 1989; 40:33–39.

34. Tupker RA, Pinnagoda J, Coenraads PJ, Nater PJ. Susceptibility to irritants: Role of barrier function, skin dryness, and history of atopic dermatitis. Br J Dermatol 1990; 123:199–205.

35. Van der Valk PGM, Kruis-de Vries MH, Nater JP, et al. Eczematous (irritant and allergic) reactions of the skin and barrier function as determined by water vapour loss. Clin Exp Dermatol 1985; 10:185.

36. Potts RO. Stratum corneum hydration: Experimental techniques and interpretation of results. J Soc Cosmet Chem 1986; 37:9–33.

37. Salter DC. Further hardware and measurement approaches for studying water in the stratum corneum. In: Elsner P, Berardesca E, Maibach HI, eds. Bioengineering and the Skin: Water and the Stratum Corneum. Boca Raton, FL: CRC Press, 1994:205–215.

38. Walling PL, Dabney JM. Moisture in skin by near infrared reflectance spectroscopy. J Soc Cosmet Chem 1989; 40:151–171.

39. Salter DC. Instrumental methods of assessing skin moisturization. Cosmet Toiletr 1987; 102:103–109.

40. Murahata RI, Hing SAO, Maibach HI, Roheim JR. Use of a microwave probe to evaluate the hydration of human stratum corneum in vivo. Bioeng Skin 1986; 2:235–247.

41. Leveque J-L, de Rigal J. Impedance methods for studying skin moisturization. J Soc Cosmet Chem 34:419–428, 1983.

42. Salter DC, Hodgson RF, Hall LD, et al. Moisturization processes in living human skin studied by magnetic resonance imaging microscopy. J Invest Dermatol 1992; 100:529.

43. Querleux B, Richard S, Bittoun J, et al. in vivo quantification of the mobile water content in skin layers by high spatial resolution MR imag-

ing. Proceedings of the Society for Magnetic Resonance in Medicine, 12th Annual Scientific Meeting, 1993:943.

44. Willis CM, Stephens CJM, Wilkinson JD. Assessment of erythema in irritant contact dermatitis. Contact Derm 1988; 18:138–142.

45. Crowe DM, Willard MS, Murahata RI. Quantitation of erythema using reflectance spectroscopy. J Soc Cosmet Chem 1987; 38:451–455.

46. Crowe DM, Willard MS, Murahata RI, et al. Objective measurement of surfactant irritation by fiber optic spectroscopy. Contact Derm 1988; 19:192–194.

47. Wyszecki G, Stilles WS. Color Science: Concepts and Methods. Quantitative Data and Formula, 2d ed. New York: Wiley, 1982:165–168.

48. Nilsson GE, Otto U, Wahlberg JE. Assessment of skin irritancy in man by laser-Doppler flowmetry. Contact Derm 1982; 8:401–406.

49. Quinn AG, McLelland J, Essex T, Farr PM. Measurement of cutaneous inflammatory reactions using a scanning laser-Doppler velocimeter. Br J Dermatol 1991;125:30–37.

50. Grove GL, Grove MJ. Objective methods for assessing skin surface topography noninvasively. In: Leveque JL, ed. Cutaneous Investigations in Health and Disease. New York: Marcel Dekker, 1989:1–32.

51. Wortzman MS, Scott RA, Wong PS, et al. Soap and detergent bar rinsability. J Soc Cosmet Chem 1986; 37:89–97.

52. Jungerman E. In: Spitz L, ed. Soap Technology for the 1990s. Champaign, IL: American Oil Chemists Society, 1990:230–243.

53. Ananthapadmanabhan KP, Yu KK, Lei X, Aronson MP. On the use of fluorescein as a probe to monitor surfactant rinsability from skin. J Soc Cosmet Chem 1997, in press.

54. Finkey MB, Wagner P, Murahata RI. Evaluation of the effectiveness of a bar soap as a hand cleanser for contact lens users. J Am Optom Assoc 1988; 3:56–59.

54. Leveque JL, Grove GL, de Rigal J, et al. Biophysical characterization of dry facial skin. J Soc Cosmet Chem 1987; 83:171–177.

55. Leveque JL. Physical methods for skin investigation. Int J Dermatol 1983; 22:368–375.

9
In Vitro Testing Models for Claims Substantiation

Rolf Mast
International Beauty Design, Riverside, California

Steven Rachui
Stephens and Associates, Carrolton, Texas

INTRODUCTION

The driving force behind the explosion of in vitro science in the last 10–15 years has been the concern for animal welfare. The concept of reduction, replacement, and refinement was conceived and developed by W. M. S. Russel and R. L. Burch (1959) in their book *The Principles of Humane Experimental Technique*. Reduction means reducing the numbers of animals used without compromising results. *Reduction* lowers the number of animals involved by intelligent consideration of the known facts, including in vitro test results. It is a principle well followed in practice. For instance, Procter & Gamble reports a reduction of over 90% in the number of animals they use. *Replacement* refers to substitution, either wholly or in part, of the animal model with a nonanimal model. This includes substitution with knowledge bases and in vitro tests or sometimes with an animal lower on the taxonomic scale. *Refinement* includes techniques that can reduce or eliminate pain in the animal model. Examples are use of the low-volume Draize eye irritation tests and changing the LD_{50} to measure onset of toxicity rather than animal death, as well as limiting dose level to eliminate sheer volume effects.

"No animal testing" has become an important claim for both cosmetic and household products. It is widely used as an aggressive marketing tool by many companies. Activist groups, such as the "greens," etc., have pushed pharmaceutical and cosmetic companies to abolish or at least reduce all types of animal testing. This has become practical only since the development of the new technologies discussed in this chapter.

The September 1995 edition of *Scientific American*, the 150th anniversary issue, is dedicated to the prediction of key technologies for the twenty-first century. Within this journal, L'Oreal has printed a two-page account of their vision for the future. Their 3D in vitro skin model is the major featured technology. They feel that the use of reconstructed skin to study penetration, metabolic changes of applied substances, aging, photoprotection, etc., represents a major opportunity for cosmetic technology in the twenty-first century. It would not be amiss for the cosmetic industry to share the same idea, especially as extended to in vitro technologies in general.

In the main, the in vitro technologies discussed in this chapter were developed for product safety reasons. However, when attributes like aging, photoprotection, and penetration can be studied with those techniques, it is fairly clear that methods originally developed for nonanimal safety work will become a major part of our arsenal for cosmetic performance evaluation. This is especially true as more and more functionality is being built into products. The rise of the "cosmeceutical" category has been precipitated largely by three factors: the explosive growth of the alpha hydroxy acid products with concomitant demand by consumers for more effective products; the increased realization in the cosmetic industry of the inherent bioactivity of the skin and the consequent use of functional products to mediate skin bioactivity; and the rise of in vitro testing technology to provide a practical means to evaluate product functionality with respect to skin damage and metabolism.

Another part of product claim work is, in fact, product mildness. Anyone who has run an extensive consumer test will recognize that even products that are clinically mild and do not elicit overt irritation can elicit a negative response from 5–15% of consumers on subjective feelings of stinging and burning. In vitro tests can differentiate mildness between otherwise acceptable products. This can lead to improved consumer satisfaction and claims. Further, the in vitro methods themselves are measures of bioactivity and thus are very adaptable to functional claims in general.

It might be noted that as discussed here, and generally, in vitro testing means the characterization of biological effects without the use of potentially pain-sensing organisms. Over the last decade, there has been an enormous growth in the development of in vitro methods. Most of this effort is related to correlating the tests with in vivo toxicological studies, with much effort being directed at replacing the Draize eye irritation method. To provide a broader picture, therefore, in vitro tests developed as animal alternatives, especially those of foremost concern to the cosmetic chemist, are given a general description below.

The bulk of this chapter is divided into two segments: (1) a discussion of the various new in vitro methods that have been developed and (2) examples and case studies on the utilization of these methods for both claim substantiation and product development. Given that in vitro methods were developed in the context of safety testing, it is best to discuss these methods functionally from that context. This is the subject matter covered below. The text further deals more directly with how in vitro techniques can be used for the purpose of general claim development.

DESCRIPTION OF SOME MAJOR IN VITRO METHODS

An overview of different in vitro test types serves as a background to viewing them for formula development and claim substantiation. Bear in mind, however, that their basic function during development was for animal safety test replacements; thus much of the description relates to safety issues. We have chosen to group them into the following seven categories, each of which has different applications and attributes. A pictorial representation of these systems is shown in Figure 1.

The following seven technologies are discussed:

1. Cell culture/cytotoxicity tests
2. Artificially grown skin models

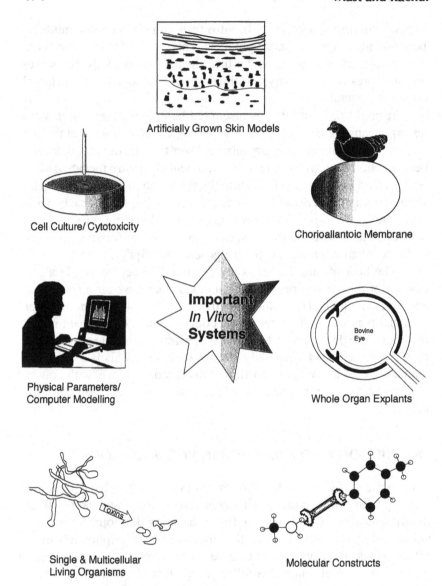

Figure 1 Some major in vitro testing methods.

3. Molecular constructs
4. Physical parameters and computer modeling
5. Single-cell and multicellular organisms
6. Whole organ explants
7. Chorioallantoic membrane (CAM) assays

Cell Culture/Cytotoxicity Tests

General

Relevant cells such as human keratinocytes can be cultured and used in relatively simple screening studies. There is a strong correlation between whole-body and cell-culture toxicity. On the other hand, there is no stratum corneum barrier to moderate toxic effects, and, of course, no morphological, immunological, or exact cell match with living systems. Another drawback is that substances must be soluble in the applied aqueous medium. This eliminates many insoluble and solid materials.

Cultures of individual cells are grown and tested for viability or stress after administration of the test substance. All cells operate through similar metabolic pathways and require an intact membrane for viability. Thus, in general, substances toxic to an organism are toxic to an individual cell.

Two commonly used cell cultures in cytotoxicity testing are fibroblasts and keratinocytes. These are important to the cosmetic chemist because they are, respectively, the basic dermal and epidermal cell types. Fibroblasts are found in connective tissue throughout the body. They are essential constituents in dermal structure, where they secrete a nonrigid extracellular matrix that is rich in type I and/or type III collagen. The epidermis is a multilayered epithelium composed largely of keratinocytes (so called because their characteristic differentiated activity is keratin synthesis).

Endpoints Used

Because all cells are metabolically related, similar tests for metabolic rate and cell death can be used for different cell cultures.

Dyes. Cells can absorb certain dyes as a result of their metabolic functions. This allows widely usable test kits to be prepared, where the effect of irritants on cell cultures is measured by measuring the decreasing amount of dyestuff that is adsorbed or altered by the culture as the cells undergo toxic stress.

Two commonly used marker dye systems are the neutral red assay originally developed by Borenfreund et al. (1) and the MTT assay.

Neutral red (3- amino-7-dimethylamino- 2-methylphenazine hydrochloride, mol wt 288.8) is a water-soluble dye that passes through the intact plasma membrane and becomes concentrated in lysosomes of viable cells. Total neutral red uptake is proportional to the number of viable cells in a culture. If cell membranes and lysosomes are damaged by a test substance, dye uptake ceases; thus measurement of reduction of neutral red dye uptake (NRU) caused by a test material is an indirect measure of the material's toxicity (2).

In the neutral red uptake test, the cell culture is exposed to low amounts of irritants over a long period of time, and diminishing capacity for dye uptake is used as a function of cell death. An interesting variant of this assay is the neutral red release (NRR) test. Here healthy cells that have preabsorbed the dye are exposed to large amounts of irritant over a short time period. Now the effect on cell death and membrane rupture can be measured by dye release. This serves as a better model for short exposure to relatively corrosive agents like irritating shampoos.

The other common dye system used to determine cell viability after exposure to various test materials is MTT: 3-(4,5-dimethylthiazol-2-yl)-2,5-diphenyltetrazolium bromide). In viable cells, MTT is converted by reduction from its initial yellow color to a dark purple color by the mitochondria (the large energy-converting organelles within the cell). This purple dye, which is the insoluble dye formazin, remains internal to the cell after conversion and is later extracted with an isopropanol or DMSO wash. The intensity of the extracted purple color, as measured spectrophotometrically, is directly proportional to the metabolic activity of the cells and inversely proportional to the toxicity of the test material (Refs. 3 and 4, as quoted from Ref. 5). MTT is not compatible with all test materials. Strong reducers, for example, will stimulate the conversion of MTT and may lead to falsely elevated viability readings. Although this can be controlled, neutral red may be a better choice in some cases.

A less common but potentially more efficient dye is Alamar blue. Unlike MTT and neutral red, which require extraction for effective viability measurements, Alamar blue does not alter cell structure and does not require extraction. This allows the same tissue to be used for viability and histology measurements.

Other Cytotoxicity Endpoints.   Other cytotoxicity endpoints, such as the measurement of pH (i.e. Silicon Microphysiometer), can also be used. Typically, pH changes of 10–100% occur over a time scale of seconds to minutes as cells are contacted with a test substance. These measurable effects are related to changes in cell viability due to the test substance (6–8).

Fluorescein leakage is used as an indicator of membrane integrity, especially through initially coherent layers of corneal epithelial cell cultures. Damage caused by test substances can disrupt the membrane. This disruption, and hence the potential irritancy of the substance, can be measured by fluorescein leakage through the cell structure. In the bovine corneal opacity and permeability (BCOP), test, opacity and permeability are measured. Permeability is affected by things like surfactants while opacity may not be affected.

Other important methods include protein synthesis and cytokine evaluation under different cell stresses. The latter are cell mediators involved in both irritant and allergic skin inflammations. They are produced by the human keratinocyte cells grown either in cell cultures or differentiated skin models. Measurement of the mediators expressed with very mild trauma is forming the basis of extremely sensitive analysis of the effect of cosmetics on the skin. Cytokines can be released by irradiation as well as contact with chemicals, so they can be used to study the effects of UV exposure and sunscreen amelioration of damage on tissue. In this context Corsini et al. have examined the effects of sunscreen on the release of interleukin-1 and tumor necrosis factor-alpha (TNF-α) from human keratinocytes exposed to UV irradiation (9).

Artificially Grown Skin Models

These provide a surprisingly good morphological and histochemical match with natural epidermis. In the MatTek epidermal model, for instance, the lipid profile is almost identical to human skin, and its stratum corneum is similar in structure and permeability to living human skin. In addition to cytotoxicity, enzyme expression and inflammatory mediators from the keratinocytic cells can be measured. However, many structures such as blood vessels, lymph node, Langerhans cells, etc., are missing, which eliminates the natural skin allergic reactions of edema and swelling. The stratum corneum in artificial skin can be more permeable than in living skin, which, in turn, can affect the results obtained.

Unlike medium-covered cell cultures that are composed of uniform functional cells, cultured skin equivalents have a differentiated structure that resembles human epidermis. They are formed by allowing human keratinocytes to be exposed to air as they are cultured. This causes the keratinocytes to flatten into a surface layer closely resembling the stratum corneum. A major advantage of these systems is that the effects of solids and water-insoluble materials can be conveniently determined because direct contact with the dried surface is possible.

The in vitro skin equivalents all have a differentiated epidermal structure grown from keratinocytes, but the models differ depending upon the underlying structure. The latter can range from being a cultured dermal model to a mechanical substrate with no dermal properties. Prunie'ras has defined the possibilities in some detail (10).

An epidermal model (EpiDerm), and a skin-equivalent model (Skin^2) are discussed below.

EpiDerm

The EpiDerm system is available from the MatTek Corporation, Ashland, MA. It is a multilayered, highly differentiated epidermal model grown from neonatal foreskin-derived human epidermal keratinocytes. They are cultured using serum-free media in Millicell CM (Millipore Corp.) cell culture inserts. EpiDerm consists of organized basal, spinous (prickle), granular, and cornified (squamous) layers, as exist in human epidermis. The progression of keratinocytes into a stratum corneum similar to the in vivo situation is demonstrated by the presence of keratohyalin granules, tonofilament bundles, desmosomes, and a multilayered stratum corneum containing intercellular lamellar lipid layers arranged in patterns characteristic of the in vivo epidermis.

Both cytotoxicity and irritation endpoints are possible, including the MTT, interleukin-1 alpha (IL-1 alpha), PGE$_2$, and LDH assays, as well as sodium fluorescein permeability. The prostaglandin PGE_2 is a member of the eicosanoid group, which are synthesized by all cells. Prostaglandin extracellular chemical mediators, expression of which increases with cell damage, are involved in the in vivo inflammation response, making them good markers for irritant reactions. Interleukins, such as IL-1 alpha, are released by cells to mediate the immune response of helper T cells. Lactate dehydrogenase (LDH) is an enzyme universally found in cell cytoplasm. Release of all these mediators occurs with cell damage. Metabolically significant cell mediation indicators are especially sensitive because they can measure pretoxic irritant effects. Artificial endpoints such as the MTT assay and fluorescein permeability are more measurements of cytotoxicity.

Of particular interest to the cosmetic chemist is the lipid analysis in the EpiDerm stratum corneum layer and the diffusion of water through the structure (11). Lipid intercellular lamellar sheets with broad-narrow-broad spacing similar to in vivo stratum corneum have been observed. Furthermore, chemical analysis shows similar concentrations of lipids and ceramids as found in normal skin. In initial tritiated water-diffusion experiments, both EpiDerm types, EPI-100, and the more mature EPI-612 showed similar but higher water permeability coefficients than cadaver skin (5.0×10^{-3}, 3.6×10^{-3}, and 2.5×10^{-3} cm/h respectively.)

The utility of the EpiDerm model in ranking shampoo irritation has been clearly demonstrated (12).

Skin² Models

Advance Tissue Sciences (La Jolla, CA) had made several Skin² models. A detailed overview has been published (13) but unfortunately these models are no longer available. The technology, however, is still very relevant and is reported in this article as information. Note that similar materials, which are still available from other sources, can be used in its place for very similar effects.

Modeled in the ZK 1300 system are a dermis, epidermis, and stratum corneum composed of metabolically and mitotically active cells. The dermal layer consists of neonatal foreskin fibroblasts cultured onto a nylon mesh, resembling human skin in gross structure (14). Close parallels also exist in the immunochemical examination of the $Skin^2$ model

(15), which looked for specific proteins in the model compared to in vivo human skin. Epidermal differentiation was confirmed by the presence of K1-keratin, trichohyalin, and filaggrin. Extensive correlation to dermal protein was also noted.

Ceramide lipids, important to the barrier function, have been shown by thin-layer chromatography (TLC). The permeability of radiolabeled water was similar to that of neonatal foreskin. Also, the model metabolized testosterone to a profile of metabolites similar to that of neonatal foreskin. The researchers believe that the in vitro skin model will be useful for the study of drug permeability and metabolism (16).

In addition to morphological structures and protein/lipid compositions, metabolic functions have also been correlated (13). This includes lipid synthesis and enzyme activities comparable to in vivo epidermis.

Because of ZK 1300's close similarity to actual skin structure, it can be used to test any product in physical form (liquids, solids, lotions etc.) for a variety of claim substantiation issues including moisturization sun protection factor (SPF), phototoxicity, collagen formation, irritation, etc.

Relatively simple measurements can be made through cell viability assays using MTT uptake. However, due to the metabolic activity of the differentiated keratinocytes, irritant response can also be measured through a variety of proinflammatory mediators, interleukins, and enzyme endpoints (17).

Another use for Skin^2 models is to examine the effects of phototoxic chemicals as evidenced by the use of ZK 1351, which is a fibroblast keratinocyte model (18). The investigators showed 90% correlation to existing in vivo data.

Molecular Constructs

By their very nature, these approaches can only provide empirical data. They are designed to mimic the physical interactions between an irritant and an organism and can provide good screening data.

As defined in this chapter, molecular constructs are in vitro constructs that do not use living or excised cells. From a toxicological point of view, the molecular construct reacts with test materials to produce endpoints that can be empirically related to in vivo results. In effect, a standard curve is produced that can be used to assess the irritation of unknown test systems provided they are related to the test systems that were used to produce that curve.

Very well characterized molecular constructs are available from In Vitro International and are discussed immediately below. In a more general sense, molecular constructs can be any documented or ad hoc benchtop system that allows an initial screening of test material physical properties so final in vivo testing can be limited to those most likely to succeed. Possibilities here are endless. A few examples are given later for illustrative purposes.

In Vitro International EYTEX, SKINTEX, and SOLATEX Systems

The scientific basis of the EYTEX system has been described in detail by Gordon (19). At heart it is an ordered protein gel whose subunits are a purified Jack Bean protein of 30,000 mol wt. This transparent network also includes smaller proteins, peptides, amino acids, aminoglycans, and mucopolysaccharides that contribute to the overall response of the irritant. This matrix is constituted in a cuvette, with an inert partition membrane at the top surface. Passive diffusion of the test sample through the membrane into the matrix causes turbidity, which can be measured. The EYTEX model is constructed so that the measured turbidity correlates with Draize in vivo eye irritancy. Other proprietary ingredients called "enhancers" and "stabilizers" are included. Not surprisingly, modifications in the test have evolved to accommodate different classes of materials (20), so that the empirical correlations become more and more accurate within those classes. Table 1 summarizes these tests.

The SKINTEX method has the same core technology as the EYTEX system but is, of course, designed to mimic results of the rabbit or human skin irritation test. Here a buffered salt solution of keratin, collagen, and a dye is bound to cellulose within a plastic disk that is superimposed upon the ordered macromolecular matrix. This extra biobarrier was designed to assess the effects of known chemicals that bind to or alter the stratum corneum. Changes in the integrity of the barrier cause a dye indicator to be released. In addition, changes in the matrix can alter its transparency. These parameters are measured together spec-

Table 1   In Vitro International EYTEX Assays

Test Designation	Test Material Class
Upright membrane assay (UMA)	A broad screening protocol
Rapid membrane assay (RMA)	Primarily utilized in screening surfactants
Alkaline membrane assay (AMA)	Specifically used for alkaline materials
High-sensitivity assay (HSA)	Used to detect variances in low irritation samples

trophotometrically. Again, the basic system has been modified to fit different exigencies, as shown in Table 2.

Similar systems related to the measurement of UV damage and protection are sold under the SOLATEX name.

Some other possible ideas for claim substantiation using the SKINTEX/EYTEX systems or other types of molecular constructs are listed below:

Measure rinsability of surfactant formulas. Thus, after application, what amount of washing is needed to avoid a subsequent irritant endpoint? Similarly, how do physical characteristics affect the rate at which irritation is produced?

Measure efficacy of cationic deposition from different anionic surfactant systems. Cationics left after washing would cause measurable irritancy.

Table 2   In Vitro International SKINTEX Assays

Test Designation	Test Material Class
Upright membrane assay (UMA)	A broad screening protocol
High sensitivity assay (HSA)	Detects subtle irritation differences in low irritation samples
Alkaline membrane assay (AMA)	Used for alkaline materials
Human response assay (HRA)	Designed to predict human in vivo response
Standard labeling protocol (SLP)	Used to predict 4-h rabbit Draize skin test

Incorporate different anti-irritants into the test formula or the molecular construct directly. This will help determine mechanism and efficacy (i.e., is the effect to limit absorption or to form a less irritating complex?). How does counterirritancy depend upon anti-irritant molecular weight? etc.

Use as a quick screen of SPF characteristics within a defined system. This could act as a quality-assurance protocol during product manufacture.

Recently, modifications to the Eyetex and Skintex systems have been made. The new test system called "Irritection" is based on a 96-well microtiter plate format. Although the basic idea behind the test is consistent, these modifications are designed to make the system more accurate, more reproducible, and less labor-intensive.

Other Possible Constructs

The area of molecular constructs can be broadened to include any number of physical systems that will show a response to different formulations.

Some examples are to study the effects of surfactants on protein agglomeration as an indicator of irritation. Other examples of simple molecular constructs that need validation follow-up but have been used with some success are as follows:

1. A product to be tested for occlusivity was coated onto individual, tightly curled hair fibers in a humidity chamber. Where more occlusive films were used, those based on petrolatum, the fiber dropped out much more slowly than where less occlusive films based on mineral oil were used.

2. The efficacy of bovine serum- or silicate-based temporary wrinkle removers can be measured using a rubber membrane attached to a balance. The skin-stretching ability of antiwrinkle products can be estimated by measuring changes in scale weight as the experimental mixture dries.

3. In vivo living skin can be used as a "molecular construct" by treating it with anionic surfactant and then measuring the absorption of that surfactant by treating the skin with a poorly absorbed cationic dye. Staining is directly related to anionic surfactant absorption. This allows

the effects of different anti-irritants to be determined very quickly if they work by reducing surfactant uptake on the skin.

4. A membrane-covered UV-sensitive polymerization reaction could predict the effect of sunscreens coated on the membrane as the apparatus is left exposed to the sunlight. These are the acrylic-based systems routinely used in UV-cured inks, etc. The effectiveness of the sunscreen would be proportional to the delay it caused in effecting the polymerization reaction. Potentially this system would be quite effective in measuring different interesting parameters. For instance, UVA protection could be measured by using UVA radiation–potentiated photoinitiators; similarly with UVB radiation. Another possibility is measuring the effect of different waterproofing systems by washing the test material–coated membrane in different ways

These and an infinite number of other possible constructs are not proof of in vivo performance. On the other hand, they are very quick and inexpensive. They can be used on an ad hoc basis by the formulator to greatly increase the chances of success with a final formula, thus allowing the elucidation of interesting new ideas and interactions.

Physical Parameters and Computer Modeling

These relate measurable physical parameters to toxicity potential. One interesting sidelight of using these theoretical techniques is that they are, of necessity, related to molecular structure considerations. This can be useful in stimulating formula development.

Computer Modeling of Structure Versus Activity

These studies fall in the general area of quantitative structure activity relationships (QSAR). This can very much relate to performance as well as safety issues.

Certain aspects of irritation and/or allergic interaction with the skin depend upon measurable physical parameters. Included in this are molecular weight, ionization, hydrogen bonding potential, polarizability, pH, and partition coefficient. A discussion of these factors as they relate to skin permeation, etc., is beyond the scope of this chapter; however, some interesting articles have been published relating physical parameters to both in vivo skin irritation and sensitization in a systematic manner.

Sigman et al. have developed a CADES (Contact Allergens Data Evaluation System) computer model that includes 1200 substances tested as primary or cross sensitizers (21). These are related to structural and chemical properties databases. In addition to acting as a very useful resource for looking up known data, CADES can be used to help evaluate new and untested agents. Structure-activity analysis with CADES may help clarify mechanisms of sensitization and develop chemicals with reduced allergenicity potential.

Modeling of allergic contact dermatitis can be done even more directly (22). In this computer model, known allergens and nonallergens are carefully analyzed in terms of a series of physical parameters such as partition coefficient, hydrophilic/lipophilic structures, polarizability, H-bonding potential, selected radical substructures, etc. Using an algorithm based on these physical parameters, 30 of 38 known allergens were correctly classified, while 37 of 42 known nonallergens were correctly classified. The authors did point out that the failure to correctly classify urushiol, a known potent allergen, does mean that additional refinements are clearly needed. Nevertheless the very encouraging degree of correlation does make this a potentially useful in vitro technique for preliminary screening.

Consideration of Basic Physical Parameters

When large-scale consumer trials of almost any skin cosmetic are run and respondents are carefully questioned about whether they experienced any irritation, there is a positive response of reported irritation (stinging, redness, etc.) that usually runs into the 5–10% level. Mostly these are subjective irritation complaints, not serious dermatological conditions. Elimination of these low-level interactions with the skin is an important formulation expertise and could have a big influence on ultimate product success in the marketplace.

In certain circumstances, knowledge of physical parameters can help a formulator to increase the probability of making a milder product that is less likely to produce this background level of irritation complaints. Some examples are to use nonpenetrating polymeric emulsifiers in creams and lotions instead of surfactants that can penetrate and irritate the skin. Also, for shampoos and soaps, the use of anti-irritants (such as nonionics and amphoterics) that can interact with the primary anionics in the formula or the actual ethoxylation of the anionics themselves can change skin absorption and irritation properties. Staying with surfactants, it is well documented that aromatic structures such as cetylbenzalkonium chloride and nonylphenoxy nonionics are more irritating than their alkyl counterparts. Other obvious issues to consider are pH, molecular size, and lipid solvent power.

Single and Multicellular Living Organisms

Given that we need not worry about the suffering of single-cell organisms or even simple multicellular species and that these can be dealt with at a "test-tube" level, these systems can also be included in the overall category of in vitro test systems. This is not covered here except to say that one widely reported system, which has much literature correlation with in vivo irritation tests, is the Luminescent Bacteria Toxicity Test. It uses a marine bacteria *Photobacterium phosphoreum*. Respiratory metabolism in this bacterium is indirectly coupled to the light output caused by the activity of the bacterial luciferase enzyme in cellular oxidation reactions. Microbics Corp., of Carlsbad, CA, has developed a commercial toxicity screen based upon this reaction, sold as the Microtox system. It measures viability of *P. phosphoreum* through photometrically measured light output. The endpoint of the assay is the concentration of test material that decreases light output by 50%. The effect of an irritant to kill the bacteria can thus be conveniently measured in a kit form suitable to laboratories in general. The test does suffer similar disadvantages to the cytotoxicity tests above, including the need for soluble reactants devoid of preservatives, thus limiting the test's ability to measure many fully formulated systems.

This test system is widely used in water testing.

Whole-Organ Explants

Whole-organ explants in the context of this chapter are excised eyes and skin. They offer the obvious advantage of using fully formed tissue relevant to the reaction being studied. Test materials can be individual chemicals as well as fully formulated, solid, water-soluble and water-insoluble products. The need for fresh samples of the organs in question, however, makes these methods generally less convenient. Despite the added inconvenience associated with obtaining fresh organs, many of these methods correlate well to in vivo data. The bovine corneal opacity and permeability test, as developed by Gautheron, has been used to evaluate personal care products and eye area cosmetics. The data obtained with the bovine cornea correlate well with existing in vivo data for these same compounds (5). Other researchers have reported excellent results (23,24).

Chicken Enucleated Eye Test
One example by Dutch workers discusses the chicken enucleated eye test (CEET). This measures the effect of irritants on tissue closely resembling the in vivo situation. In this case, chicken eyes freshly collected from a slaughterhouse were treated with the potential irritant and examined under a microscope for corneal swelling, corneal opacity, and fluorescein retention. A comparative study with 21 chemicals showed that the chicken EET was very reliable and accurate in assessing the eye irritation potential of test materials without the use of laboratory animals (23).

Bovine Eye Models
More commonly reported in the literature than the use of chicken eyes is the use of bovine eyes. Measuring the effect of chemicals on extracted cow eye lenses is an interesting technique because of the possibility of measuring reversible eye damage in vitro (25). Thus the lens can be kept functional in a growth medium for extended periods and its condition measured optically. Reversibility of ocular damage is an important aspect of live animal testing that has been difficult to measure in other ways. Since the lens is maintained in an intact state in fluids that imitate those inside the eye, it retains its normal recuperative powers and can be used to measure recovery from damage.

A simpler technique is to measure corneal opacity induced by a test substance in an excised bovine cornea. A comprehensive paper shows the effect of 44 personal care products and cosmetics on bovine eye corneal opacity and permeability compared to the same materials tested on the Skin^2 model ZK1200 (5). Corneal opacities were read using an opacitometer; permeability was measured using fluorescein leakage from an anterior to a posterior compartment of the cornea holder. Individual scores in the bovine system correlated to animal data 25 times out of 28, for an accuracy of 89%.

For formula development, eye explant technology is an attractive possibility for studying the irritating effects of materials designed for use in the eye area under highly exaggerated conditions. Products in question include facial cleansers, facial sunscreens, facial masks, wrinkle removers, and eye makeup/makeup removers. For example, if new solvent systems for the removal of waterproof mascara were being evaluated, an eye transplant would be a good way of greatly reducing the possibility of the remover causing corneal opacity. In addition to avoiding use of animals during the formulation development, tests of this type would allow much greater flexibility in exposure and subsequent rinseoff than the formal rabbit eye Draize test. Also, the possible seepage of product under contact lenses could be simulated, as could the deliberate introduction of eye makeup solids and abrasion. Another possible use is to prescreen products that are deliberately introduced into the eye. Two possibilities here are to prescreen (at higher concentrations) materials that are to be used in human eye sting tests, to ensure no long-term opacification, as well as to evaluate eyewash products. Another potential use is that as novel aqueous based hair sprays are introduced the effect of aerosol contamination into the eye could be checked. For instance, it is possible that some water-based latex systems would be more damaging than the solvent systems currently in use.

Excised Human Skin

No attempt will be made to review this extensive literature except to mention the following recent work because of the interesting tie-ins to subjects discussed elsewhere in this chapter.

An interesting method originates from the TNO Toxicology and Nutrition Institute (P.O. Box 360, 3700 AJ Zeist, Netherlands) (23). They have developed a two-compartment skin organ culture model in which either human, rabbit, or porcine skin remains viable for several days (26). The epidermal side remains free of direct contact with the culture medium and retains the stratum corneum as a physiological barrier. Test substances can be applied directly to the skin surface in concentrations similar to in vivo exposure. Both water-soluble and water-insoluble materials can be evaluated. Two quite different effects of irritants on the skin culture can be measured. These are cytotoxicity based on the MTT assay as well as measurements of arachidonic acid derived inflammatory mediators—in particular 12-hydroxyeicosatetranoic acid (12-HETE) a proinflammatory mediator, and 15-HETE an anti-inflammatory mediator. It was also noted in results with different levels of citric acid that the skin cultures can return to normal MTT conversion after initial damage, so the cultures can be used to measure recovery.

Chorioallantoic Membrane (CAM) Assays

It should be noted that the discussion of this assay is adapted from Bruner (Ref. 27, pages 163–165).

The chorioallantoic membrane (CAM) is the vascularized respiratory membrane that surrounds the embryonic bird within an egg. The idea is to evaluate the toxicity of a test material as a function of changes produced in the CAM.

To run this assay, the egg is incubated for 3 days at 37°C and a hole is then cut to expose and isolate the CAM at the "equator" of the egg. The window is then resealed and incubation continued for another 14 days to allow the CAM to develop fully. Test materials can then be added to the CAM, incubation continued for an additional 3 days, and test material–induced lesions measured. The commonly used CAMVA assay is similar, but the lesions on the CAM are scored at just 30 min after treatment by the test material.

In the related HET-CAM assay (i.e., hen's egg test), eggs are incubated whole, large end up, for 10 days, and the shell is cut away at the air sac to expose the CAM. Test substance application is for 20 s, followed by warm-water irrigation.

A significant benefit of the CAM test is that it can be used to test virtually any water-soluble or insoluble raw material or product. Water-soluble materials are dosed with or without dilution, while hydrophobic materials may be dissolved into vehicles such as mineral oil before being applied directly to the surface of the CAM (28).

DISCUSSION OF IN VITRO TESTS IN THE CONTEXT OF PRODUCT DEVELOPMENT

The previous section discussed technologies that mimic the in vivo situation mainly with the intent of making predictions about product safety. This opens the door to all sorts of possibilities to adapting these methods in the development of new products and supporting the claims made for these products. Some of the advantages of in vitro claim substantiation techniques include protocols that allow novel and very subtle effects to be measured, some mechanistic analysis to be made, and the possibility of obtaining a very quick result. Further, techniques that would be impractical, too dangerous, or unethical on human beings or animals might well be reconstructed using in vitro substrates because they allow testing with no need for safety data. For these reasons they have become an important research tool.

Some possibilities in this regard are discussed below, as follows:

1. Differentiation of supermild products
2. Evaluation of anti-irritant action
4. SPF and UV exposure
5. Antiaging/antioxidants
6. Antiaging/wrinkle formation
7. Collagen synthesis
8. Wound healing
9. DNA synthesis
10. Keratinocyte turnover rate
11. Evaluation of unknown materials
12. Radioactive studies and inconvenient protocols
13. Lipid and moisturization studies

Differentiation of Supermild Products

It is important for the formulation chemist to differentiate between degrees of mildness, not just to measure irritation. Defining formulations that are too irritating to market is a relatively straightforward concept. On the other hand, all cosmetics elicit varying degrees of low-level skin reactions, which are becoming more and more important to identify and minimize. Thus, when consumer tests are run on products that have been approved for irritancy using the Draize rabbit skin irritation test, they can still elicit a subjective irritation response from the consumers using them, often in the 5–15% reaction complaint range. If reformulation is necessary, this can be a very expensive and time-consuming misfortune. However, if in vitro mildness studies to determine the mildest formula variants are run prior to the consumer test, one would have an excellent chance of not running into subjective irritation problems. Further, as the marketplace and one's competitors become increasingly sophisticated in these issues, reducing subjective irritation will become more and more important to stay competitive.

This section briefly mentions several ways in which in vitro methods are being used to determine the mildest variant of otherwise acceptable products.

Some new techniques for mildness discrimination are outlined in Table 3.

Irritation causes cell death, which can be measured in cell cultures using dye marker techniques. The key to mildness measurement, however, is the measurement of cell mediators expressed prior to cytotoxicity. The commercially available ocular and skin models can be utilized easily to measure the leakage of enzymes associated with cytotoxicity, mediators of inflammation (prostaglandins), and mediators associated with the complete immune response (interleukins). Based on study design, these endpoints can be used to discriminate "mildness" among products formulated to be nonirritants.

Using artificial skin models (see above), researchers have demonstrated a decrease in the release of the inflammation mediator prostaglandin (PGE_2) with a hypoallergenic versus a regular marketed

Table 3 Some "Mildness" Evaluation Techniques

Protocol	Provider
Skin^2 model	Advanced Tissue Sciences
EpiDerm and EpiOcular models	MatTek Corporation
High sensitivity assay	In Vitro International
Bovine cornea mildness assay	Stephens & Associates
Epithelial cell permeation	Microbiological Associates

skin lotion. Furthermore, the results are obtained much more quickly than with patch testing. Thus, of the three mild lotions tested, one formulation was noted as causing more PGE_2 release than the other formulations after 24 h of exposure. The same mildness ranking was shown with patch testing, but this took 14 days (29).

In related work, Jackson et al. studied the difference in IL-1 alpha release caused by three hypoallergenic moisturizers in a skin model system (30). The data again correlated very well with the 14-day cumulative irritation index measured in vivo. Given that all three products were hypoallergenic formulations, the living skin model gave a quick result and predicted subtle differences between formulations scoring zero in normal patch testing. In vivo sensitization potential of the products was measured using a 14-day induction phase with challenge testing after 2 weeks. None of the products were sensitizing, but some cumulative irritation was observed. This was compared to cytokine expression in a product-treated skin model (Skin^2) measured 24 and 48 h after treatment. Components released from the treated cells—lactate dehydrogenase (LDH), prostaglandin E_2 (PGE_2), and IL-1 alpha were measured. The results, shown in Table 4 for IL-1 alpha, are taken from their paper. The results show that IL-1 alpha can be used to differentiate between mild formulations.

It is well known that consumer response to product mildness involves totally subjective stinging reactions that cannot be measured as well as the sorts of conditions (like redness) that can be measured and/or judged externally. Currently, stinging reactions are gauged by applying product to sensitive people in the nasal fold area, which is an awkward and subjective test. It may be possible to use skin models to correlate

Table 4   Cytokine Expression Versus Skin Irritation in the Skin^2 Model

Product	In Vivo Irritation (14 days)	IL-1-alpha pg/ml (24 h)
Product X	23	66.98
Product Y	5	26.68
Product Z	73	115.27
Pos. control	318	—
Neg. control	—	47.3

mediator release of prostaglandins and related leukotrienes with stinging reactions. Such an objective measurement could help in the development of non itch/sting products.

EpiDerm is an artificially grown skin model from the MatTek Corporation. It may be used in a similar manner to discriminate differences in mildness between products. This model is different in that it consists only of epidermal keratinocytes and a stratum corneum with a barrier function and lipid profile closely approximating that of human skin. This difference does not compromise the quality of mediator and dye-based (MTT) cytotoxicity assays (see above). An advantage of this model is that it is relatively inexpensive

There is a new method to discriminate mildness between test materials. This model utilizes 8-mm bovine cornea sections dosed topically for 24 h. At the end of exposure, tissue viability is confirmed with MTT, and the levels of prostaglandin mediator (PGE_2) released are measured (using an enzyme-linked immunosorbent assay kit from PerSeptive Diagnostics). This PGE_2 measurement is particularly sensitive. This model uses real corneal tissue and not a synthetic construct. Further, the experiment uses a replication set of $n = 6$ for statistical evaluation. Other common models use replications of only $n = 2$ or 3 because of the associated cost.

In a totally different model, Virginia Gordon at In Vitro International has said that the SKINTEX high sensitivity assay (see above) was developed to measure the mild 0–2 region of the 0–6 Frosch/Kligman irritation assay. This makes it very useful for mildness development work. For instance, it is used for alpha hydroxy acid formulations, where it can pick out milder products even at the same pH and AHA concentration. Work has also been done with the optimization of preservatives and preservative levels versus the irritation they elicit in skin products using this same technique.

One group has been working with epithelial cell permeation techniques originated at the Medical College of Pennsylvania (31). Here, epithelial cells are grown to form a solid layer across the bottom of a transwell. The layer does not allow passage of a fluorescein market dye because of the tight junctions formed between the cells. This barrier is very similar to that found in the cornea. When dosed with an irritating material, these junctions break down and the marker dye is able to pass.

This technique has also discriminated between otherwise very mild products that had all scored zero on in vivo studies. It is applicable to creams and lotions, etc., as well as water-soluble surfactants and sounds like a great method to help eliminate consumer reaction complaint surprises.

Overall, there are now a number of very interesting techniques for discriminating between mild and supermild products.

Evaluation of Anti-Irritant Action

Formulators more and more are attempting to develop the most skin-compatible products possible. How well, in fact, do materials like sodium glycyrrhizinate, bisabalol, etc., work as anti-irritants?

In vitro methods can be used to discriminate fine degrees of irritation, as just discussed. They also can be used to evaluate the possible effects of anti-irritants used in cosmetics. This not only at a level where obvious irritation is suppressed but also at a level where the anti-irritant could lessen the more subtle effects of consumer-perceived cosmetic reactions. Obviously, the same techniques would be used, as just discussed, for mildness, but rather than examine different (but related) products, the same formula would be evaluated for effects on leakage or mediator release in the presence or absence of an anti-irritant. It might be mentioned that MTT is a good marker in this context. By measuring cell half-life with or without an anti-irritant in the formula, it is possible to determine statistically measurable differences.

A specific example where this has been successfully implemented is in the use of the SKINTEX high-sensitivity assay with AHA formulations. The AHA irritation was plotted against concentration of different added anti-irritants. A smooth curve defining the optimum anti-irritant level can be elicited. A technique like this could be very useful when one considers the rate at which anti-irritant technology is expanding. For AHAs alone, the following have been listed: bisabolol, green tea extract, aloe vera, kola nut extract, sphingosine, beta glucans, oil of rosemary, and bois oil (32). A method that might rank these and others could be very useful.

Work has also been done with shampoo anti-irritants using this same technique.

In Vitro Systems and UV Exposure

Moving to sun care, in vitro methodologies provide new tools for measuring SPF, waterproofing, and phototoxicity, which will help both with development and quality-control work. More importantly, the techniques are able to develop information relating to the antiaging aspects of skin care.

Some of the in vitro methods used fall into the categories shown in Table 5.

A few comments on SPF techniques are relevant here.

In Vitro SPF Techniques

The use of in vitro models to measure SPF has been reported using the Skin^2 model. The skin model allows easy adaptation to measure water resistance, which could make it a very convenient adjunct to the formulation work.

The SKINTEX model has also been adapted to measure SPF in the SOLATEX SPF system, discussed above. Here the macromolecular matrix has been treated with UVB-sensitive materials that cause turbidity in the matrix when exposed to UV light. Using known sunscreens, an exact response versus SPF curve has been produced. Further, the pro-

Table 5 Some In Vitro Methods for UV Exposure

In Vitro Class	Some Measurements Taken with UV Exposure
Cell cultures	Photoirritancy—neutral red uptake
	Cytokine expression
	Viability/UVA exposure
	SPF/tanning
Artificially grown skin models	Photoirritancy—MTT measurement
	SPF
	Cytokine expression
	Antioxidants/anti-irritants
Molecular constructs	Photoirritancy
	SPF
	QA and anti-irritancy

tocol is easily adapted to study water resistance. Using this technique, the effects of formulation on SPF can be quickly and conveniently measured. Again, in addition to formulation work, it could also be used for quality control, to validate proper dispersion of TiO_2 formulation.

Gordon et al. presented a detailed evaluation of SOLATEX SPF in a three-laboratory SPF study of 100 formulations (33). There was a 84–96% interlaboratory concordance and a linear correlation coefficient of $r = 0.92$ to $r = 0.96$ for 53 to 57 sunscreen formulations. A useful feature is the easy adaptation of in vitro systems to different methodologies, such as a waterproofing protocol. Again, the authors consider the method to be a useful and fast screen prior to in vivo evaluation.

Cell cultures can be used to measure SPF effects that are not possible with simple physical systems. In this context, an imaginative use of keratinocytes to substantiate the effectiveness of self-tanning products was conducted by Croda (34). The Diffey-Robson technique is an in vitro measurement of UV absorption using tape and a spectrophotometer. Self-tanning products that darken skin and thus help with the sun protection cannot be measured by the standard Diffey-Robson technique simply because they do need skin to interact with to produce their effect. One example is juglone (5-hydroxy-1,4-naphthaquinone), which interacts with skin protein to form colored compounds with UV absorption properties. In order to estimate the UV absorption effects of juglone on skin, Gallagher used human keratinocyte cultures in place of the normal tape substrate of the Diffey-Robson method. The juglone was able to interact with the keratinocytes, darken them, and the film could then be assessed for UV transmittance.

Antiaging

Antioxidants

It is well known that one of the detrimental effects of UV exposure on the skin is the production of free radicals, with the concomitant metabolic breakdown leading to erythema and cell death. The use of antioxidants increases the effective SPF of products by interfering in this process. Skin models can be used to directly study the effect of antioxidants in the prevention of skin damage caused by solar radiation.

Thus some very interesting work using the (now unavailable) Living Skin Equivalent (LSE) has been done (35). The work was based on observations that UV radiation causes an increase in lipid peroxides, followed by a decrease in enzymatic and nonenzymatic antioxidant defense mechanisms in human epidermal skin. These changes have been recorded in vitro using LSE, which is a viable organotypic model of human skin containing a derm-like fibroblast/collagen matrix, a stratified epidermal component and a stratum corneum. LSEs were exposed to UVB ($100-1500$ mJ/cm^2) or UVA ($1-10$ J/cm^2) and then assayed for hydroperoxides by a hemoglobin-catalyzed reaction in which a colorless methylene blue derivative is converted to methylene blue, which can be measured spectrophotometrically at 674 nm. An increase in color formation was observed in a dose dependent manner for both UVB and UVA irradiated LSEs. Work showed that pretreating the skin with an antioxidant inhibited color formation. This indicated that UV exposure does result in formation of reactive oxygen species and that applying antioxidants to the skin is an effective countermeasure. These data also show that in vitro skin models can be used for studying free radical mechanisms in skin. The fact that both UVA and UVB effects were measured is of significance in the increasing desire to formulate for antiaging as well as sunburn protection factors.

The relative antioxidant efficacy of different materials has more recently been evaluated using the MatTek EPI-100 epidermal model, discussed above (36). The technique was quite different from that just quoted, because the effect of antioxidants on cytokine expression was used to indirectly measure lipid peroxides. The prostaglandin E_2 (PGE_2) was measured as a marker for antioxidant activity. The literature records that an increase in oxygen radicals will lead to an increase in lipid peroxides, which leads to an increase in PGE_2. A total of 13 materials were evaluated for antioxidant properties. Results showed that this method is a useful tool for screening antioxidants. The following list of materials was evaluated providing useful data for the formulator of skin protectant cosmetics: mixed tocopherols, water-soluble liquorice extract, quercetin, potassium glycyrrhinate, delta tocopherol, actiphyte of white grape, magnesium ascorbyl phosphate, liposomal superoxide dismutase, green tea extracts, and an experimental formula.

This method accurately predicted that both mixed and delta tocopherol as well as liposomal superoxide dismutase were efficacious antioxidants. The efficacy of these materials is well documented. Data are shown in Figure 2.

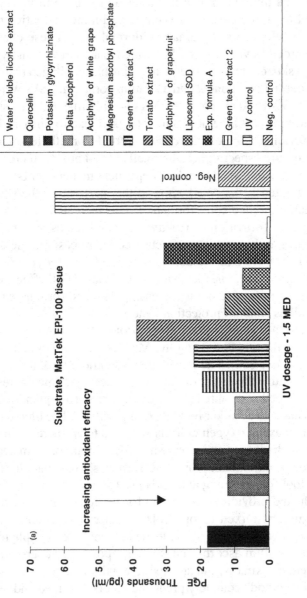

Figure 2(a) Epidermal model, antioxidant screening assay (36).

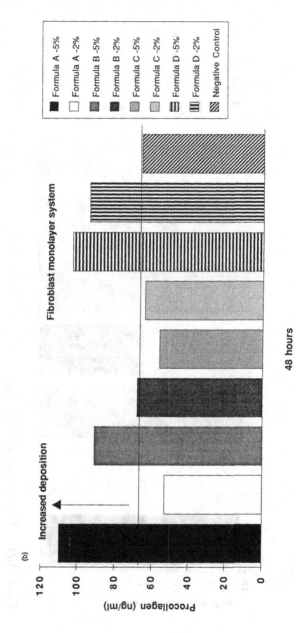

Figure 2(b)   Epidermal model, antioxidant screening assay (36).

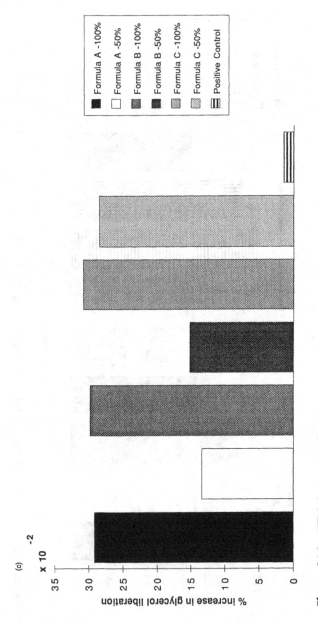

Figure 2(c) "Thigh cream" efficacy assay.

Wrinkle Formation

Prevention of wrinkles is the major issue driving skin care cosmetic use. It is not hard to measure the short-term amelioration of wrinkles facilitated by skin creams. On the other hand, from an ethical point of view one certainly would not want to deliberately cause wrinkles in subjects through UV exposure in any sort of controlled test, to say nothing of the inconvenience of the protocol. In vitro models that can systematically predict the effect of UV radiation on wrinkles and thus measure cosmetics to diminish the effect are of great potential significance.

Two papers leading in this direction are as follows:

Biederman has reported on work done by Procter & Gamble, where cytokine expression was measured following UVA exposure of in vitro human cells (37). They found that human cells irradiated with a single dose of UVA secreted specific cytokines (IL-1B and IL-6) that subsequently upregulated the expression of several proteases, including elastase and collagenase 1. These proteases are known to mediate skin matrix damage; thus their increased expression can be tied to chronic events leading to skin aging. This work was related to in vivo expressed elastase from chronically UVA-exposed mouse skin. This induction occurred earlier than visible expression of skin sagging in mice. In addition it was found that UVA exposure produces chromosome aberrations that are associated with carcinogenesis.

The effects of UVA and UVB radiation have been compared on fibroblasts and on hairless mouse skin (38). Increased fibroblast cell viability in the presence of a UVA absorber can be demonstrated. In parallel experiments, UVA radiation was shown to increase wrinkles in hairless mice, which were ameliorated in the presence of UVA absorbers. Kim suggested that the in vitro method might be useful for evaluating sunscreen materials and testing phototoxicity. Mice, however, would be better suited to evaluate the materials effective in interrupting the harmful effects of free radicals generated by UV radiation than the fibroblast cell cultures.

In Vitro Systems and the Optimization of OTC Drug and Cosmetic Activity

The metabolic functions of cell cultures, skin equivalents, etc., open up a rich mine of possibilities for the study of the metabolically related effects of OTC drugs as well as activity related to the use of cosmetics. It also allows for studies that are difficult with in vivo methods, such as radioactive techniques and evaluation of unknown materials. Without getting into a debate on what the cosmetic industry should and should not claim, this section discusses a few possibilities in this area.

Of course, the methods in this section do not model the in vivo situation in an exact scientific sense for the obvious reasons. The idea is that work using the in vitro models will at least aid the formulator to screen formulations that are more likely to be effective in any definitive in vivo trial. The formulator using these methods will also need to work around the practical difficulties that ensue. Cell cultures, for instance, cannot be exposed to real use concentrations and formulations. The effect of pH in assessing effects of AHAs must be considered, etc. In vitro methods are not in any sense "turnkey" operations. An intelligent use of what is taking place is necessary, with the limitations needing clear insight and understanding.

Collagen Synthesis

Collagen synthesis can be studied using fibroblast monolayers or fibroblast containing skin models.

Collagen synthesis may be measured through the use of staining techniques (aniline blue), isotope assays (tritiated proline), or by measuring procollagen. The amount of procollagen present is directly proportional to the amount of collagen synthesized. Thus procollagen measurements are a direct measure of collagen deposition.

Monolayer studies are cost-effective and quick but have the disadvantage of not allowing neat test materials to be assayed and not having the epidermis and stratum corneum present. However, monolayer models can be useful in evaluating the activity of formulations and raw materials. Typical data obtained with these systems are shown in Figure 2b.

Skin model ZK1301 is representative of full-thickness human skin models which allow the use of neat test material. The studies conducted with this model were generally conducted as rapidly as with monolayer tests but were more labor intensive. Using the proline assay, the levels of collagen types I, II, and III synthesized in the ZK 1301 have been reported to increase (39).

Because of the sensitivity of these models and the dosing periods required by these assays, the materials used should be noncytotoxic. A cytotoxic agent may cause cell death prior to the formation of collagen. Because materials may be cytotoxic in vitro after such a long exposure but not be cytotoxic in vivo, a cytotoxic material may be judged non-efficacious in vitro even though it is effective in vivo. The user of in vitro systems for collagen deposition must make sure that sufficient controls are included to ensure the accuracy of the data generated.

Wound Healing

Obviously any in vitro model is far removed from the complex in vivo wound healing process. Nevertheless, a preliminary evaluation of the effects of potential healing agents on damaged keratinocytes based skin models could be useful and does provide a basis for controlled bench-top studies.

Zeigler et al. have examined aspects of the Skin[2] model as it relates to wound healing (40). Using very sophisticated immunological tech-niques, the effect of overlying keratinocytes on underlying fibroblast extracellular matrix formulation (ECM) was examined. It was found that fibronectin formulation in the dermal equivalent layer was dramat-ically increased by the presence of an overlying keratinocyte layer. These increases in fibronectin deposition, as measured by immunofluo-rescence, were due to increases in steady-state mRNA levels in resident fibroblasts. In turn, this was stimulated by epidermal growth factor (EGF) and transforming growth factor beta (TGF-β) produced by ke-ratinocytes. Fibronectin has discrete binding regions for collagens and glycosaminoglycans and facilitates the organization of the ECM. Other interactions between dermal and epidermal layers were discussed. Potentially, this type of methodology provides a way of determining the effect of materials on skin regeneration (rejuvenation) at a detailed mol-ecular level.

In other experiments, a simple blade cut is made in the Skin[2] model surface and the rate of reepithelialization is measured. Surpris-ingly, the keratinocyte model can respond in a manner related to in vivo wound healing (39).

Presumably, other skin models could be used in similar ways.

DNA Synthesis and Inhibition

In addition to collagen, Rogers is using cell cultures to measure the effect of different materials on DNA synthesis or inhibition which, of course, is one measure of the overall rate of cell turnover (41). One application of this work could be the evaluation of new drugs for the prevention of hyperkeratinization problems such as psoriasis, and it could be very useful in developing a quality-control method for coal tar extracts as well as optimizing OTC-based formulations in general. Outside of the personal care industry, this could be a screen for the evaluation of totally new drugs for potential use in diseases relating to keratinocyte hyperproliferation.

Keratinocyte Turnover Rate

Addition of chemicals to growing living skin equivalents might help to expand cosmetic science. Lotions that allowed formation of a well-structured, orderly skin might be a lot more preferable for everyday use than lotions that distorted or inhibited growth. Looked at slightly differently, this would be a way to directly measure the effect of ingredients on the rate of skin formation. These techniques might, for instance, help to quell the worries that the FDA has about the long-term use of AHAs. The same would apply to other so-called cosmeceuticals like the "slimming creams." Also, the work would be complimentary to that done on the effects of chemicals on in vivo tape-stripped epidermis where the effect of different chemicals on the restructuring skin growth can be measured.

Radioactive Studies

The feasibility of doing radio-tagged studies with in vitro models opens up other interesting possibilities in such areas as penetration studies, barrier functions, and the orientation of moisturizing molecules.

Examples of the types of questions that could be answered are as follows:

1. How much do liposomes help penetration of active materials through the skin?
2. What is the effectiveness of penetration enhancers such as Penederm, oleic acid, and azone?

3. Evaluation of barrier creams and cosmetics that stop the penetration of irritants such as anionic and cationic surfactants through the skin. For instance, applying a high-molecular-weight polypeptide, with many basic groups, to the skin could bind anionics and stop further damage.

4. How do externally applied claim ingredients incorporate themselves into skin structure? For instance, how do externally applied sphingolipids or other lipids end up incorporated into the intercellular lipid structure? Do larger molecules such as "soluble collagen" and mucopolysaccharides penetrate into the skin? Further studies are needed of the interaction between lipids and humectants and on cell turnover and growth as they relate to AHA concentration, etc.

Lipid and Moisturization Studies

EpiDerm and Skin2 models differentiate into a stratum corneum–like layer, with well-defined and closely related lipid bilayers between the flattened cells. This means that they have potential for a study of the effects of ingredients on formation of the lipid layer and possibly measurement of absolute moisturization in the skin.

Metabolism of Excipients

Both skin models and two-compartment skin explants have been used to study the metabolic breakdown products of applied materials. This is because keratinocyte systems have extracellular expression of enzymes similar to that which occurs in vivo. This process is critical for the effective utilization of active precursors like vitamin E acetate, and retinol palmitate.

Thigh Creams/In Vitro Tests with Nondermal Tissue

A method has been introduced to measure the efficacy of "thigh creams" using in vitro methodology. The premise of this method is that thigh cream materials will break down triglycerides to free fatty acids and glycerol. The glycerol liberated from the fat tissue is quantified spectrophotometrically (42). The relative efficacy of thigh cream formulations may be measured by comparing their ability to cause the release of glycerol from the adipose tissue. Typical data are shown in Figure 2c.

CONCLUSION

Over the last several years it has been increasingly realized that cosmetics can have numerous effects on the complex biochemistry of the skin. This has coincided with rapid advances in the development of in vitro techniques, primarily for the purpose of assessing product safety. It is not surprising, therefore, that these two issues have merged to where in vitro techniques have been adapted to measure performance issues beyond product safety. Specific performance effects of raw materials can be more easily predicted and formulation of superior products made possible. This is greatly contributing to the growth and vitality of the cosmetic industry.

REFERENCES

1. Borenfreund E, Puerner J. Toxicity determined in vitro by morphological alterations and neutral red absorption. Toxicol Lett 1985; 24:119–124.
2. Clonetics publication. Multiple applications of the neutral red bioassay. Cell Commun 1989; 1(3):
3. Carmichael J, Degraff WG, Gazdar AF, et al. Evaluation of a tetrazolium based semiautomated colorimetric assay: Assessment of chromosensitivity testing. Cancer Res 1987; 47:936–942.
4. Triglia D, Braa SS, Donnelley T, et al. A three-dimensional human dermal model substrate for in vitro toxicological studies. In: Goldberg AM, ed. In Vitro Toxicology: Mechanisms and New Technology. Vol 8. New York. Mary Ann Liebert, 1991:351–362.
5. Rachui SR, Robertson WD, Duke MA, et al. Predicting the ocular irritation potential of cosmetics and personal care products using two in vitro models. In Vitro Toxicol 1994; 7(1):45–52.
6. McConnel HM, Owicki JC, Parce JW, et al. The cytosensor microphysiometer: Biological applications of silicone technology. Science 1992; 257:906–1912.
7. Parce JW, Owicki JC, Kercso KM, et al. Detection of cell-affecting agents with a silicone biosensor. Science 1989; 246:181–296.

8. Bruner LH, Miller KR, Owicki JC, et al. Testing ocular irritancy in vitro with the silicone microphysiometer. In Vitro Toxicol 1991; 5:277–284.

9. Corsni E, Marnovich M, Galli CL. Effects of different sunscreens on ultraviolet radiation-induced cytokine release by a human keratinocyte cell line, as reported by Wolven Garrett A. Drug Cosmet Ind 1994; June:12–14.

10. Prunie'ras M. Skin and epidermal equivalents: A review. In: Rougier A, Goldberg, AM, Maibach HI, eds. In Vitro Skin Toxicology. New York: Mary Ann Liebert, 1994:97–105.

11. EpiDerm. Brochure from MatTek Corporation, 200 Homer Ave, Ashland, MA 01721.

12. Doyle JM, Dressler WE, Rachui SR, Stephens TJ. Evaluation of two in vitro skin equivalents (the EpiDerm and Skin[2] Model ZK1300) for assessing the skin irritation potential of personal care products and chemicals. From MatTek Corp, 200 Homer Ave, Ashland, MA.

13. De Wever B, Rheins LA. Skin[2]; An in vitro human skin analog. In: Rougier A, Goldberg AM, Maibach HI, eds. In Vitro Skin Toxicology. New York: Mary Ann Liebert, 1994:121–129.

14. Conrad P, Bartel RL, Jacobs L, et al. Culturing keratinocytes and fibroblasts in a three-dimensional mesh results in epidermal differentiation and formation of a basal lamina-anchoring zone. J Invest Dermatol 1993; 100(1):35–39.

15. Fleischmajer R, MacDonald ED II, Contard P, Perlish J. Immunochemistry of a keratinocyte-fibroblast co-culture model for reconstruction of human skin. J Histochem Cytochem 1993; 41:1359–1366.

16. Slivka SR, Landeen LK, Zeigler F, et al. Characterization, barrier function, and drug metabolism of an in vitro skin model. J Invest Dermatol 1993; 100(1):40–46.

17. Slivka SR, Zeigler F. Use of an in vitro skin model for determining epidermal and dermal contributions to irritant responses. J Toxicol Cutan Ocul Toxicol 1993; 12:49–57.

18. Edwards SM, Rheins LA, Sayre RM. A novel in vitro assay for assessing phototoxicity using a human skin model, as reported by Wolven Garrett A. Drug Cosmet Ind 1994, June 12–14.

19. Gordon VC. The scientific basis of the Eytex system. ATLA 1992; 20:537–548.

20. Gordon VC. Utilization of in vitro technologies in cosmetic formulation. May 25, 1993. Obtained from In-Vitro International, Irvine, CA.

21. Sigman CC, Bagheri D, Maibach HI. Approaches to structure-activity relationships in skin sensitization using CADES. In: Rougier A, Gold-

berg AM, Maibach HI, eds. In Vitro Skin Toxicology. New York: Mary Ann Liebert, 1994:271–280.

22. Magee PS, Hostynek JJ, Maibach HI. Modeling allergic contact dermatitis. In: Rougier A, Goldberg AM, Maibach HI, eds. In Vitro Skin Toxicology. New York: Mary Ann Liebert, 1994:281–291.

23. TNO Toxicology and Nutrition Institute, P.O. Box 360, 3700 AJ Zeist, Netherlands. Toxicology Tribune no. 13, 1993.

24. Prinsen MK, Koeter HBWM. Justification of the enucleated eye test with eyes of slaughterhouse animals as an alternative to the Draize eye irritation test with rabbits. Food Chem Toxicol 1993; 31(1):69–76.

25. Sivak JG. An alternative to Draize test using lenses from abattoir supplied eyes of cattle: Trends and alternatives in testing. Joseph F. Morgan Res Found Newslet 1994; spring/summer:6–8.

26. Rutten, AAJJL, Bequet-Passelecq, Koeter HBBWM. Two compartment model for rabbit skin organ culture. In Vitro Cell Dev Biol 1990; 25:353–360.

27. Bruner LH. Ocular irritation. In: Frazier JM, ed. In Vitro Toxicity Testing: Applications to Safety Evaluation. New York: Marcel Dekker, 1992:149–190.

28. Goldberg AM, Silber PM. Status of in vitro ocular irritation testing. Lens Eye Tox Res 1992; 9(3&4):161–192.

29. Rheins LA. Man made skin can be used to test possible phototoxic drugs. Dermatol Times 1994; 15(6).

30. Jackson EM, Stephens TJ, Rheins LA. Assessing hypoallergenic facial moisturizers using in vivo and in vitro tests. Cosmet Toiletr 1994.

31. Harbell JW. A competitive edge for in vitro product development and safety testing. Presented to the CTFA 51st Scientific Conference, Oct 25, 1994.

32. Goldenberg RL. Drug Cosmet Ind 1995; Jan: 37.

33. Gordon VC, Hamilton D, Boone S. The SOLATEX system to measure photoirritation and sun protection factors. Presented to IFSCC, September 1993.

34. Gallagher KF. Self-tanning the walnut way. Soap Perf Cosmet 1994; March:43.

35. Pelle E, Mammone ET, Combatti M, et al. (abstr 641). J Invest Dermatol 1993; 100(4):

36. Rachui SR, Robertson WD, Duke MA, Allen R. An in vitro method to screen antioxidant efficacy. Communication from Thomas J. Stevens and Associates, Inc.

37. Biederman KA. Identification of UVA markers of skin damage. Annual

Scientific Seminary Society of Cosmetic Chemists, Cleveland Ohio, May 4–5, 1995.

38. Kim EJ. 18th International IFSCC Congress, Venice, Oct. 3–6, 1994, as reported in Soap Cosmet Chem Spec 1994; Nov:64.
39. Communication with L. Rheins. Adv Tiss Sci (12/94).
40. Zeigler FC, Landeen L, Naughton GK, Slivka SR. Tissue-engineered, three dimensional human dermis to study extracellular matrix formation in wound healing. J Toxicol Cut Ocul Toxicol 1993; 12:303–312.
41. Communication with T. Rogers, Southern Research Institute (12/94).
42. Rachui SR. Evaluation of thigh cream efficacy. Communication from Thomas J. Stevens and Associates, Inc., 1995.

10

Sensory Testing Methods for Claims Substantiation

Clare A. Dus, Marie Rudolph, and Gail Vance Civille
Sensory Spectrum, Chatham, New Jersey

INTRODUCTION

> "The plusher brush zeroes in on every lash to lengthen and thicken incredibly"
> "Without a clump in sight"
> " . . . natural looking color that lasts and lasts, up to 6 weeks"
> "Works on the surface to encourage dry skin to feel normal"
> " . . . it changes instantly into a creme that slips on without streaking"

These claims were gleaned from the advertising pages of *Mirabella* and *Ladies' Home Journal*. At first glance the language usage suggests hype and the advertiser's cleverness at hooking the consumer; however, closer inspection of the wordsmithing reveals terminology chosen to state a claim or reveal a truth about the sensory properties of the mascara, hair colorant, creme, or makeup foundation being marketed. Words such as *lengthen, thicken, clump, slip,* or *streaking* are terms based on sensory perception and performance. These words not only describe a product but may also differentiate the product from similar products, thus allowing the advertiser to use the sensory terms to convince consumers to buy a particular name brand in order to meet the consumers' sensory expectation. Manufacturers of consumer products sell sensory properties; they develop products with specific sensory properties. Therefore, it is not only appropriate but also necessary to evaluate or measure sensory properties.

Sensory testing methodology has a long history of use in the food industry. Early methods consisted of grading goods such as tea or coffee. The coffee taster or professional tea taster derive from these classification systems. Modern sensory evaluation methods were conceived during the 1930s, 1940s, and 1950s (1–3). Today sensory evaluation is an integral part of the food product development cycle; from benchtop screening to consumer testing. Many food manufacturers dedicate departments to the measurement of appearance, flavor, and texture characteristics of their products.

Formal sensory evaluation methods are a recent addition to the product development cycle within the personal care industry. In the past, personal care products were developed with input from expert evaluators.

Sensory testing methodology, developed for the evaluation of foods and beverages, has been adopted and adapted by the personal care industry. Consumer tests, difference tests, and descriptive tests are successfully utilized to evaluate the sensory characteristics of hand and body lotions, shampoos, hair mousses and gels, antiperspirants, shaving creams, mouthwashes, and toothpastes (4–6).

SENSORY PROPERTIES

Physiology

What instrument do we use for sensory measurements? The human body is host to millions of sensory receptors. For example, the optic nerves of the eye (vision), the nerve endings in the skin surface (touch), the olfactory epithelium located in the roof of the nasal cavity (olfaction), or the taste buds located on the tongue (taste) (1). These receptors are capable of perceiving a complex interaction of external stimulation and translating this into a measurable response (7,8).

Sensory responses may be subjective when they are based on consumer expectations, or sensory responses may be objective with carefully controlled test variables such as environment, sample handling, and respondent (panelist) selection and training. Differences in perception among panelists are overcome by creating a common sensory experience. The use of a frame of reference to develop a common library of terms with definitions, strict evaluation protocols, and the use of intensity scales with physical references ensures that trained panels produce valid and reliable data.

Terminology Development

What are the sensory properties that drive consumer acceptance and liking of a product? When consumers rate a lotion or cream as very *creamy*, what do they really mean? Descriptive sensory evaluation provides a means of describing products through the use of clearly defined terminology. Early training of panelists with respect to the broad physical and/or chemical principles that underlie sensory properties—such as the visual, tactile, or force/stress related factors involved in skin-feel properties—is an important first step. Armed with this basic knowledge, panelists are introduced to a large array of samples from the product category of interest. Using standard evaluation procedures, the panelists develop a list of terms that best describe the characteristics of the product category (9).

Several descriptive terms have a commonality across applications as illustrated in Table 1. These terms all relate to a common definition or description but require a different procedure based on the sample type. In foods, the firmness of a product is evaluated during the initial bite, while in skin care, firmness of a sample is evaluated as it is compressed between the index finger and thumb. In both cases it is the force required to attain a given deformation that is being measured.

Terminology for Personal Care Products

Perceived product performance and overall liking of products in the cosmetic and personal care industries are greatly influenced by their sensory qualities. Through the use of sophisticated sensory descriptive and consumer techniques and statistical analyses, the sensory attributes that drive consumer liking and perceived performance can be identified. Using common terminology for a product type with descriptive testing

Table 1   Texture Terms

Hardness: The force to attain a given deformation
  Foods ⟹ Firmness, hardness
  Skin care ⟹ Firmness, force to spread
  Fabric ⟹ Force to compress, force to stretch
Adhesiveness: The force required to remove a sample from a given surface
  Foods ⟹ Sticky (tooth/palate), toothpack
  Skin care ⟹ Sticky/tacky, drag
  Fabric ⟹ Hand friction (drag), fabric-to-fabric friction
Moisture release: The amount of wetness/oiliness exuded
  Foods ⟹ Initial moistness, juiciness
  Skin care ⟹ Wetness
  Fabric ⟹ Moisture release

provides unique and comprehensive profiles by which samples can be compared and the true differences identified. Products within a specific category will have some specific terminology (lotions and creams—pickup attributes such as firmness and peaking; antiperspirants and deodorants—application attributes such as force to apply and ease of spreading; and liquid, bar, and gel soaps—application attributes such as bubble size and rate of lathering). They also share common terms such as *residue* (amount and type) and *stickiness*. Industry has begun to realize the need for a unified lexicon to improve communication within a company and among suppliers and customers to measure product changes carefully so as to improve development timing and success rates. Many follow the guidelines set forth by the American Society for Testing and Materials (ASTM) for several product evaluation categories. Examples of descriptive terms used in the personal care, cosmetic and fragrance industries are given in Tables 2 through 4. Within each of these product types, the sensory cues will be visual (appearance), olfactory (fragrance/smell) and/or tactile (texture/feel).

SENSORY TEST METHODS

The nature of the claim determines the sensory test method. Suppose the claim is "Our product is the same as theirs." An overall difference test will provide the information necessary to make the claim. If the claim is "No shampoo is as thick as our shampoo," an attribute difference test is the appropriate test. Finally, if the claim is "Our facial moisturizer is preferred over brand X," then a consumer test is the correct choice among the sensory test methods. Sensory methods are discussed in detail in several publications (1,10,11) and in relationship to advertising claims in an ASTM document entitled "New Standard Guide for Sensory Claim Substantiation" (12).

Discrimination

Discrimination tests include difference tests and descriptive tests. Difference tests measure possible sensory differences within a set of samples and are well documented in the literature (1,10,11) These tests are categorized into two groups: overall difference tests and attribute difference tests. Each group of tests is designed to answer specific questions. Overall difference tests answer the question "Does a sensory difference exist between samples?" and attribute difference tests answer the question "How does attribute X differ between samples?" The advantage discrimination tests offer over consumer tests is that valuable information about products is obtained with usually less than 50 subjects under controlled conditions. This makes discrimination testing economically attractive compared to consumer tests.

Overall Difference Tests

Triangle Test.   This test consists of three samples. Two samples are identical, one is different (Fig. 1). Subjects are asked to identify the odd or different sample among the three. Analysis of the results of triangle tests is based on the number of correct answers compared to the number expected by chance alone when no differences exist. Statistical significance is determined by use of published tables (1). The nature of the triangle test is such that one would expect the odd sample to be selected by chance one-third of the time. It is recommended that 20–40 subjects are used for this test. The triangle test has limited use with products that generate sensory fatigue, carryover, or adaptation.

Duo-Trio Test.   This test also consists of three samples; two samples are exactly alike, one is different. However, in this case, one of

Table 2 Terminology for the Sensory Properties of Personal Care Products

Lotions and creams	Antiperspirants	Soaps	Shampoos and conditioners
Appearance	Appearance	Appearance	Appearance
Color	Color	Color	Color
Hue	Hue	Hue	Hue
Chroma	Chroma	Chroma	Chroma
Intensity	Intensity	Intensity	Intensity
Gloss	Opacity	Opacity	Opacity
Opacity			
Integrity of shape	Application	Application	Application
Roughness (lumpy, grainy)	Ease of application	Rate of lather	Ease of dispensing
	Ease of spreading	Thickness of lather	Rate of lather
Rubout		Amount of lather	Amount of lather
Wetness	Afterfeel	Bubble size	Rinsability
Spreadability	Coolness		
Thickness	Gloss	Afterfeel (before dry)	Afterfeel (wet)
Absorbency	Amount of residue	Rinsability	Combability
	Stickiness	Slipperiness	Wetness

Afterfeel
 Gloss
 Stickiness
 Slipperiness
 Amount of residue
 Type of residue
  Oily
  Waxy
  Greasy

Pickup
 Firmness
 Stickiness
 Cohesiveness
 Peaking

Wetness
Occlusion
Tautness
Whitening

Amount of residue
Type of residue
 Soap film
 Oil
 Wax
 Grease

Afterfeel (after dry)
 Stickiness
 Slipperiness
 Amount of residue

Coldness
Amount of residue
Detangling
Stickiness of hair

Afterfeel (dry)
 Combability
 Amount of residue
 Gloss
 Slipperiness

Table 3   Terminology for the Sensory Properties of Cosmetic Products

Appearance	Application	Afterfeel
Color	Ease of application	Tautness
Intensity	Amount of coverage (%)	Slipperiness
Chroma	Streaking	Amount of residue
Gloss	Wetness/moistness	
Pearlescence		Type of residue
		Oily
	Removal	Waxy
	Ease of removal	Greasy
	Amount left on skin	Powdery

the identical samples is labeled as the "control" or reference and the subject is required to pick which of the two remaining samples is a match to the control (Fig. 2). The duo–trio test is less efficient than the triangle test because the probability of selecting the correct sample by chance is 50%.

Table 4   Basic Descriptive Fragrance/Odor Classes

Citrus	Sweet	
Fruity	Brown	
Stone	Floral	
Berry	Fruity	
Melon		
Tropical		
	Spice	Wood/resinous
	Peppery	Balsamic
Aldehydic	Brown	Pine
Ozone/marine		Camphoraceous
Woody/powdery		
	Herbaceous	
	Mint	Animal
Floral	Green	Leather
Violet	Mossy	Musk
Rose		Fecal
Muguet		
White flower		

Judge No. _____ Name _____ Date _____

Type of Sample: *Antiperspirant*

INSTRUCTIONS:

Smell the sample on the tray from left to right. Two samples are identical; one is different. Select the odd/different sample and indicate by placing an X next to the code of the sample. If you wish to comment on the reasons for your choice or if you wish to comment on product characteristics, you may do so under remarks.

Sample Code	Odd Sample	Remarks
457	[ ]	_____
390	[ ]	_____
458	[ ]	_____

Figure 1   Example of a score sheet used for triangle tests.

Simple Difference Test.   This method is used when the composition of the product or the application of the product renders it unsuitable for multiple presentation, as in the case of as mascara or facial moisturizer or shampoo (half-head tests). Each subject is presented two samples and asked whether the samples are different or the same (Fig. 3). The presentation of samples is randomized, so that half the time two different samples are presented and half the time two of the same samples are presented. The results are analyzed by using the χ^2 (chi-square) test and comparing the number of "different" responses for the matched pair (A/A and B/B) to the number of "different" responses for the different pair (A/B and B/A). The presentation of the matched pair enables

Judge No. _____ Name _____ Date _____

Type of Sample: *Antiperspirant*

INSTRUCTIONS:

Smell the sample on the tray from left to right. The left hand sample is a reference. Determine which of the remaining samples matches the reference and indicate by placing an X next to the code. If a difference is not apparent between the two unknown samples, you must guess. If you wish to comment on the reasons for your choice or if you wish to comment on product characteristics, you may do so under remarks.

Sample Code

Reference   539 [   ]   017 [   ]

Remarks _____

_____

Figure 2   Example of a score sheet used for duo–trio tests.

one to assess the influence of the placebo effect. In order to determine differences, it is recommended that 100–200 subjects be used or 20–50 presentations of each of the four sample combinations (A/A, B/B, A/B and B/A).

Similarity Testing.   Introduced by the authors of *Sensory Evaluation Techniques* (1), similarity testing is a variation on overall difference tests. Similarity tests are used when the objective is to determine similarity or no perceptible difference exists between the products thus lending itself appropriately to claim substantiation. This test method would be an appropriate choice if a claim is "Our product is the same as product X."

Any of the overall difference tests discussed here can be used as a similarity test because neither the sample preparation nor the test

Judge No. _____ Name _____ Date _____

Type of Sample: *Hair Spray*

INSTRUCTIONS:

Feel the hair swatches from left to right. Determine if samples are the same/identical or different. Mark your choice below with an X. If you wish to comment on the reasons for your choice or if you wish to comment on product characteristics, you may do so under remarks.

[   ]   Products are the same

[   ]   Products are different

Remarks: _____

_____

_____

Figure 3   Example of a score sheet used for simple difference tests.

administration changes. What does change is the statistical treatment of the data. In difference tests, the sensory analyst chooses a small α value (type I error) and assumes that there is only a small chance that samples which are not different are incorrectly declared to be different. This assumption is reversed for similarity tests. In order for the sensory analyst to be confident that samples are similar or not perceivably different, a small β value (type II error) is chosen. In this way the sensory analyst assumes there is only a small chance that different samples are incorrectly labeled to be the same. In order to account for this difference in analysis, more subjects are needed for similarity tests than for overall difference tests.

Judge No. _____ Name _____ Date _____

Type of Sample: *Treated Hair Swatch*

INSTRUCTIONS:

You have received 2 hair swatches, a control sample labeled "Control" and a test sample labeled with a 3 digit code. Feel the control sample first then feel the test sample. Indicate the size of the difference in the feel, relative to the control, on the scale below. If you wish to comment on the reasons for your choice or if you wish to comment on product characteristics, you may do so under remarks.

[　]　0 = no difference
[　]　1
[　]　2
[　]　3
[　]　4
[　]　5
[　]　6
[　]　7
[　]　8
[　]　9
[　]　10 = extreme difference

Remarks _____

_____

Figure 4   Example of a score sheet used for a difference-from-control test.

Difference from Control.   This test, sometimes referred to as a *degree of difference*, consists of a control sample and one or more test samples (Fig. 4). Subjects are asked, using an intensity scale, to rate the size of the difference between each test sample and the provided control. It is recommended that a blind control be included in the test design. Results are analyzed by analysis of variance (ANOVA).

Others. A listing and description of additional difference test methods are listed in Table 5.

Attribute Difference Tests

Attribute difference tests define the differences between two samples in terms of a specific characteristic or attribute. These tests provide more pointed information than the overall difference tests by determining specific attribute differences. The tests vary in the manner in which the questions of attribute differences are asked, ranked, or rated. The most common attribute difference tests are as follows:

Paired-Comparison Test. This test is one of the simplest and consequently the most popular test methods. It is used when the test objective is to determine whether or not two samples differ in a particular sensory characteristic (Fig. 5). Suppose a company was interested in making the claim "Our lotion is not as greasy as brand J," then this test is a suitable test to use. Before conducting a paired-comparison test, it is necessary to decide whether the alternate hypothesis is two-sided or one-sided. For claims of parity or superiority, the alternate hypothesis is one-sided. For example, the test objective may be to confirm a treatment effect (i.e., confirm that shampoo A is perceivably thicker than shampoo B). The sensory analyst counts the number of correct responses. Significance is checked in specially designed statistical tables.

Table 5 Additional Overall Difference Test Methods

Test method	Number of subjects	Premise
Two out of five	8–12	Subjects are asked to choose two out of five samples that are different from the remaining three.
"A–not A" test	30 or more	Subjects are shown examples of "A" and "not A" and then asked to determine if tests samples are "A" or "not A."

Judge No. _____ Name _____ Date _____

Type of Sample: *Hair Spray*

INSTRUCTIONS:

Feel each pair of hair swatches from left to right and mark your response below. If no difference is apparent, enter your best guess. "No Difference" responses are permitted but only as a last resort. If you wish to comment on the reasons for your choice or if you wish to comment on product characteristics, you may do so under remarks.

Test Pairs   **Which sample is *stiffer*?**

125   453   _____

874   207   _____

102   934   _____

Remarks _____

_____

Figure 5 Example of a score sheet used for a paired-comparison test.

Ranking Test. This method is used when the test objective is to rank several samples according to a single attribute—for example, greasy afterfeel (Fig. 6). Even though ranking tests are simple to conduct, one must take into account that the data are ordinal and the degree of difference is not measured. All samples, whether the differences are great or small, will be separated by one rank unit.

Rating Test. This method is used when the test objective is to rate, using some intensity scale (category, linear, or magnitude estimation), the intensity of the desired attribute (Fig. 7).

Judge No. _____ Name _____ Date _____

Type of Sample: *Hair Spray*

INSTRUCTIONS:

Feel each hair swatches from left to right and note the stiffness. Rank the samples for stiffness from least stiff to most stiff (Write "1" in the box of the sample which you find to be the least stiff, write "2" for the next stiff and "3" for the stiffest). If two samples appear to be the same, make a best guess as to their rank order. If you wish to comment on the reasons for your choice or if you wish to comment on product characteristics, you may do so under remarks.

Code:	456	871	502
Rank:	[ ]	[ ]	[ ]

Remarks _____

_____

Figure 6   Example of a score sheet used for a ranking test.

Others.   Additional test methods are listed and described in Table 6.

Descriptive Analysis Tests

Several methods of descriptive analysis are currently being used. Although the exact methodologies differ, all methods involve the detection of a series of sensory attributes and the ability to describe and quantify those attributes. The ASTM Manual Series (13) provides descriptions of the four most frequently used descriptive analysis methods [The Flavor Profile Method, The Texture Profile Method, The Quantitative Descriptive Analysis (QDA) Method, and The Spectrum Method]. A

Judge No. _____ Name _____ Date _____

Type of Sample: *Hand and Body Lotions*

INSTRUCTIONS:

Smell the lotions left to right and note the fragrance intensity. Rate each sample on the following scale:

0	none
1	
2	low
3	
4	
5	moderate
6	
7	
8	strong
9	
10	very strong

Code: 845   732   105

Rating   [   ]   [   ]   [   ]

Remarks _____

_____

Figure 7   Example of a score sheet used for a rating test.

detailed description of the Dermato Sensory Method is also available in the literature (14). Before developing or adopting a descriptive system for the purpose of evaluating and describing products, one should become familiar with the methods available and understand how each can be used to obtain meaningful data. A descriptive analysis system is selected which provides the most comprehensive, accurate, and repro-

Table 6 Additional Attribute Difference Test Methods

Test method	Number of subjects	Premise
Scheffe paired comparison	14 or more	Subjects are asked to rate an attribute of two to six samples on an intensity scale.
Rating of several samples	16 or more	Subjects are asked to rate three to six samples on an intensity scale according to one attribute. This test can be conducted as a balanced incomplete block test.
Pairwise ranking	20 or more	Subjects are asked to rank three to six samples according to intensity of one attribute.

ducible description of products and the best discrimination between them.

The basis for running a valid sensory descriptive test dictates strict adherence to protocols for panelist selection and training, test environment, sample handling, methodology, terminology development, and intensity scales with references (13).

Test Controls

The location of the panel testing area should be easily accessible to all, and it should be relatively quiet, to avoid distractions from outside conversations, machinery, and unwanted odors. Roundtables provide a means for discussion during ballot development. Testing is usually performed in booths or at tables fitted with dividers to provide privacy during the product evaluation. Sensory testing of products should be conducted under the same environmental conditions. Personal care products such as lotions, creams, and antiperspirants may exhibit differences in their sensory characteristics (i.e., absorbency, wetness, or stickiness) at different temperatures and humidities. Therefore, to claim that a product is more, less, or the same as a competitor for a given attribute or improved from an earlier formulation would not be valid if environmental test conditions were not similar. For the evaluation of fragrances, the test environment must be free from background odors and have a source of clean, odor-free air to ventilate the area. The same principle applies to lighting for appearance attributes. Do you use fluorescent, incandescent, or natural? These parameters become an integral part of the methodology employed and are followed throughout all testing. Other methodological issues to consider are the test site (place or where on the panelist), amount of product, and application procedure. This will be determined by the test objective, with input from the product developer, research scientist, market research group, and/or sensory coordinator. For claims substantiation tests, samples should be evaluated under reasonably realistic conditions.

Screening

Candidates are prescreened to determine availability, medical problems that may interfere with sensory perception, the ability to describe a sensory experience, and the ability to successfully use scalar ratings for comparison. Those who meet the prescreening criteria are then given an acuity test. This is more specific to the sensory properties and modality of perception for a product category (skin feel or fragrance of lotions and creams). The prospective panelists are tested on their ability to detect product differences both in character and intensity; their ability to describe the characteristics both verbally and using a scaling method; and their ability to relate references to various products. Candidates who have successfully completed the prescreening and acuity test phases are given an interview to assess their interest in the sensory program. They are asked questions to confirm their availability for training, practice, and workload and their communication skills and general personality. Personalities that are overbearing or extremely timid may interfere with the other panelists' ability to function as a unit.

Training

The best 12 to 15 candidates are chosen to be trained. The initial training introduces the panelists to the concept and background of sensory evaluation, scaling methods, terminology development, and how to use it all as a whole. The panelists are introduced to the scaling method that will be used. References are shown to demonstrate and anchor the perceived intensities along the scale (15). Attributes are introduced, with definitions and references. The panel practices as a whole on many different products to become comfortable with attributes and the scaling method. At first, the samples have larger differences in their sensory characteristics. As the panel progresses, the differences among products tested will be smaller. A final step to assess readiness is to validate the panel by presenting them with a set of samples, including a blind control and replicates, and review the data with respect to the panel's ability to detect and replicate intensity scores as a whole and individually. The panel is never given information about the products or the study objective. The samples are coded with random three-digit numbers. Each sample is replicated at least once. The panelists evaluate the samples in a randomized order using different orders for the first and second replications. The initial step to data manipulation is the calculation of means and standard deviations for each attribute. Usually the data are treated statistically using parametric analyses to determine significant effects. The most commonly used analysis for descriptive data is the ANOVA, utilizing the split-plot design. In this method, the sample effect is the main or whole-plot effect and judge effect and judge-by-sample interaction are the interaction or subplot effects. A multiple comparison procedure, most commonly Fisher's LSD for this type of data, is used to determine which samples differ significantly.

Affective Tests: Quantitative

Affective tests deal with consumer feelings about preference and liking and are necessary when the claim is "Our product is the preferred product" or "Our product is liked better." Quantitative affective tests answer the questions "Which product do you prefer?" and "How much do you like this product?" Unlike difference tests, affective tests are the only tests that provide an indication of how consumers feel. However, due to the inherent subjectivity of personal reaction, it is necessary to conduct affective (preference and acceptance) tests with many more subjects than difference tests. The potential influence of packaging, appearance, and/or fragrance on perceived tactile sensory or performance attributes should be minimized when possible. For parity claims, it is recommended that at least 300 consumers be used for affective tests.

Preference Tests

Preference tests asks subjects or consumers to choose one item over another (Fig. 8). The sensory analyst counts the responses for each sample and determines significance by utilizing specific statistical tables. Preference tests can be conducted as follows:

Preference Test Type	Number of Samples
Paired preference	Two
Rank preference	Three or more
Multiple paired preference	Three or more

Rank preference tests allow the consumers to indicated a relative order of preference. Subjects are asked to rank samples in order of preference with 1 = best, 2 = next best, and so forth. Attribute preference tests are

---

Name:_____   Date:_____

PLEASE SMELL THE PRODUCT ON THE LEFT FIRST. SMELL THE PRODUCT ON THE RIGHT SECOND.

NOW THAT YOU HAVE SMELLED BOTH PRODUCTS, WHICH ONE DO YOU PREFER? PLEASE CHOOSE ONE BY PLACING A CHECK IN THE APPROPRIATE BOX.

            [   ]   [   ]

            873   631

PLEASE COMMENT ON THE REASONS FOR YOUR CHOICE:

_____

_____

_____

Figure 8 Example of a score sheet used for a paired-preference test.

NAME:_____   DATE:_____

SAMPLE # 893

PLEASE LOOK AT, SMELL AND APPLY THE SUNSCREEN SAMPLE AND RATE
HOW MUCH YOU LIKE THIS SUNSCREEN FOR THE INDICATED ATTRIBUTES BY
PLACING A CIRCLE AROUND A NUMBER.

OVERALL

0	1	2	3	4	5	6	7	8	9	10
Dislike					Neither Like					Like
Extremely					nor Dislike					Extremely

FRAGRANCE

0	1	2	3	4	5	6	7	8	9	10
Dislike					Neither Like					Like
Extremely					nor Dislike					Extremely

APPEARANCE

0	1	2	3	4	5	6	7	8	9	10
Dislike					Neither Like					Like
Extremely					nor Dislike					Extremely

FEEL OF SKIN

0	1	2	3	4	5	6	7	8	9	10
Dislike					Neither Like					Like
Extremely					nor Dislike					Extremely

Figure 9 Example of a score sheet used for an acceptance test.

an option as well. Consumers are asked to indicate which sample they prefer for a given attribute; for example, "Which sample do you prefer for fragrance?"

Acceptance Tests

Use this test method when the project objective is to determine how well a product is liked by consumers. Consumers are asked, using a hedonic scale, to indicate how much they like a sample overall as well as how much they like specific attributes (Fig. 9).

Results can be analyzed using t-tests (two samples) and ANOVA (three or more samples). Attribute diagnostics can also be asked on the acceptance test questionnaire when the researcher asks consumers to rate the intensity of an attribute—for example, "How strong/intense is the fragrance of this soap?"

SUMMARY

Whether the claim is "Our shampoo leaves hair silky-soft" or "Consumers prefer our makeup remover," it will be questioned by the competition. The testing methodology will be scrutinized and the data reviewed. Therefore, it is advisable for the claimant to choose the sensory test that provides the information relevant to the objective. The testers should ask themselves, "What is the sensory claim I want to make?" and then use the sensory test that best answers the question.

REFERENCES

1. Meilgaard MC, Civille GV, Carr BT. Sensory Evaluation Techniques, 2d ed. Boca Raton, FL: CRC Press, 1991.
2. Pangborn RM. Sensory evaluation of food: A look backward and forward. Food Tech 1964; 18:1309.
3. Peryam D. Sensory evaluation—The early days. Food Tech 1990; 86: January 1990.
4. Aust LB, Oddo LP. Applications of sensory science within the personal care business: Part 1. J Sens Stud 1989; 181:
5. Oddo LP, Aust LB. Applications of sensory science within the personal care business: Part 2. J Sens Stud 1989; 187:
6. Close JA. The Concept of Sensory Quality. J Soc Cosmet Chem 1994; 45:95–107.
7. Schiffman HR. Sensation and Perception—An Integrated Approach, 2d ed. New York: Wiley, 1982.
8. Hochberg JE. Foundations of Modern Psychology Series—Perception. Englewood Cliffs, NJ: Prentice-Hall, 1965.
9. Proceedings, NAD Workshop IV, Product Performance Tests, June 1992, p 28.
10. Stone H, Sidel JL. Sensory Evaluation Practices. New York: Academic Press, 1985.

11. Amerine MA, Pangborn RM, Roessler EB. Principles of Sensory Evaluation of Food. New York: Academic Press, 1965.
12. ASTM Document—Ad Claims. In press.
13. Hootman RC, ed. ASTM Manual on Descriptive Analysis Testing for Sensory Evaluation. ASTM Manual Series: MNL 13, 1992. Philadelphia, PA 1–3.
14. Aust LB, Oddo LP, et al. The descriptive analysis of skin care products by a trained panel of judges. J Soc Cosmet Chem 1987; 38:443–449.
15. Civille GV, Dus CA. Evaluating tactile properties of skincare products: A descriptive analysis technique. Cosmet Toiletr 1991; 106(5):83.

11

Consumer Testing Statistics and Claims Substantiation

Maximo C. Gacula, Jr.
Gacula Associates, Scottsdale, Arizona

Jagbir Singh
Temple University, Philadelphia, Pennsylvania

INTRODUCTION

Claims on overall performance, liking, preference, and efficacy can be challenged through the National Advertising Division (NAD) of the Council of Better Business Bureaus (CBBB), a self-regulatory system for the industry. Consider the following example. A print advertisement for "Year 2000" body lotion came to the attention of NAD as a result of its ongoing monitoring program. The advertisement depicts a bottle of Year 2000 body lotion from company X which eclipses a bottle of company Z's "Super Lotion," positioned behind it. Company X was asked to provide substantiation for six claims, only two of which are considered below.

    1. "... try the one that makes your skin soft and smooth."

    2. "... drier and easy-to-apply ... leaves no powdery residue on your skin."

The first claim requires a consumer test and the second can be provided by a descriptive analysis technique briefly described further on. The advertiser submitted two blind central location consumer tests, the results of which are given in Table 1.

From the results of these two consumer tests, it is clear that Year 2000 was indeed preferred at a high level of confidence. This degree of confidence would be enough for a claim to be substantiated. However, upon further investigation, company X had used an older product formula marketed by company Z, which would invalidate the results of the consumer test. In this investigation, the experimental design used was found sound, reliable, and unbiased.

Two important terms used in the above inquiry need to be defined. The first is *consumer test*, defined as a method used to gather information on product performance, consumer liking, consumer attitudes and beliefs, and demographic data about a specified target population. Brief discussion of a consumer test is given in the section of this chapter entitled "Experimental Design for Claim Support." The second term is *claim substantiation*, defined simply as a statement of facts supported by evidence. This evidence is quantitative in nature and capable of being replicated under specified conditions. The quantification of claim substantiation is discussed later in this chapter.

Notice the importance of a well-designed consumer test to successfully support and/or defend a product claim. Other than the mistake of using the older product formula, the overall experimental design was sound. To avoid this mistake, it should be kept in mind that product development is a continuous improvement process and that, as a result, the marketed product formulas will be in a state of continuous change. Therefore, when planning a consumer test, one must be sure that the competitor's product to be used is current.

The evaluation of disputed claims is based on CBBB's own review

Table 1 Preference Results ($n = 200$ Respondents) for Year 2000 Lotion Versus Super Lotion

Attribute	Year 2000	Super lotion	Confidence level
Softness preference	81%	19%	99%
Smoothness preference	70%	30%	99%
Overall preference	77%	23%	99%

and, when necessary, on consultation with technical experts. Claims may be disputed for one or more of the following reasons:

1. The experiment is poorly designed to address the claim.
2. The experiment is poorly executed, i.e., protocol not strictly followed, unqualified personnel, faulty instrumentation.
3. The claim has no technical merit in relation to the product. Can the stated claim be related to product ingredients?
4. Poor choice of methods for gathering the data, i.e., instrumental methods, use of consumer and/or trained panel. Is the method used following established guidelines, such as those provided by the American Society for Testing and Materials?
5. The statistical analysis is faulty, i.e., wrong methodology used.
6. The stated claim is misleading.

STATISTICAL ASPECTS OF CLAIMS SUBSTANTIATION

Stated claims are supported by data that are statistically evaluated to judge if the experimental results are due to real effect or to random variation. Therefore, before one can proceed to develop a claim, it is important to make sure that it can be translated into a statistical research hypothesis. If a claim cannot be translated as a research hypothesis, it cannot be substantiated.

Null and Alternative Hypotheses

Briefly, a statistical hypothesis is a statement about the quantifiable aspects of products, which can be estimated from experimental results but are not otherwise directly observed. In statistical terminology, a research hypothesis (claim) is called the *alternative hypothesis*. A complementary statement to the alternative hypothesis is called the *null hypothesis*. Suppose one is testing the efficacy of product A against placebo B. From experiments for comparing A and B, we would have two sets of data, one coming from subjects treated with A and the other from subjects treated with B. If the claim is to be made that on the average, the efficacy of product A is more than that of the placebo, then statistically this is formulated as an alternative hypothesis stated as follows:

H_1: Average efficacy of A − average efficacy of B > 0

The complementary statement then becomes the null hypothesis, which is:

H_0: Average efficacy of A − average efficacy of B ≤ 0

which states that on the average the efficacy of product A is no more than that of the placebo.

The hypotheses as stated above are a comparison of the average efficacies. Again, if a claim cannot be expressed as a comparison of quantifiable aspects—as a statistical hypothesis—then the claim cannot be established. Examples of claims that cannot be translated into statistical hypotheses are statements such as "Product A is efficacious like product B," "A is a mild cosmetic like B." Such statements categorizing products do not lend themselves immediately to quantification and thus cannot be stated as statistical hypotheses for claim substantiation.

Statistical tests of significance are rules for judging whether the experimental results support the claim formulated as an alternative hypothesis. A test will either lead the investigator to conclude that there is sufficient information in the experimental data to establish the claim by accepting the alternative hypothesis or it will state that evidence in support of the claim is lacking (the null hypothesis cannot be rejected) and the claim is not established. Statistical testing will be illustrated later by an example.

Types of Errors and the Power of Statistical Test

Through experimental designs, data are collected and relevant sample statistics are computed, such as the mean, the standard deviation, etc. Since these statistics are subject to sampling and experimental errors, the statistical tests may lead to an incorrect decision. Suppose the decision is made to reject the null hypothesis and accept the alternative hypothesis. This decision, if it turns out to be wrong, is said to result in type I error. The probability of type I error is denoted by α and is known as the significance level of the statistical test. A probability of $\alpha = 0.05$ indicates that the test is liable to wrongly reject the null hypothesis 5 times in 100 cases. Significance levels $\alpha = 0.05$ and 0.01 are often used in scientific applications and are generally the accepted levels for claims substantiation. On the other hand, a type II error results if the decision is made not to reject the null hypothesis but in fact it is false. The probability of type II error is denoted by β.

In planning a claim substantiation study, both types of errors should be controlled. If, in a claim study, the consequences of type I error are serious, then we want a test procedure with a small value of α and control β as much as possible. If both types of errors are equally serious, then we may want α and β close to each other and as small as possible. Usually, the significance level α is specified during the planning stage of the claim study. Table 2 summarizes type I and II errors and their corresponding probabilities.

If H_0 is true, the probability of not rejecting it is called the *level of confidence* $(1 - \alpha)$. Similarly, if H_0 is false, the probability of rejecting it is called the *power of the test*, which is $(1 - \beta)$. Power is the ability of a test to reject H_0 when it is false. Power values range from 0 (no power) to 1 (highest power). Only tests with high power are to be relied upon in claim substantiation.

In selecting statistical tests, there is a trade-off between the significance level and the power. Given a data set, a test controlling the type I risk at a low significance level will tend to have higher type II risk and consequently low power. If an experimenter does not have the option of increasing the data size by conducting additional experimentation, then it is not possible to have a test with low significance level and desired high power. A test with significance level 0.05 will have higher power

Table 2   Types of Errors in Hypothesis Testing[a]

Decision:	If H_0 is: True	If H_0 is: False
Accept Ho	No error $(1 - \alpha)$	Type II error β
Reject Ho	Type I error α	No error $(1 - \beta)$

[a]α = probability of rejecting the null hypothesis when it is true.
β = probability of accepting the null hypothesis when it is false.

than a test with significance level 0.01. The trade-off between the power
and significance level α of a test can be seen below:

Type I error α	Confidence level	Power of test
0.01	0.99	Low
0.05	0.95	(Power increases from low to high)
0.10	0.90	
0.20	0.80	High

It is evident from that in selecting a test, one makes a compromise
between the significance level and power of the test. In consumer testing, it is a common practice to use a moderately large value of α, around
0.20, to have a test with relatively high power to detect a direction. But
this may not be a wise practice in using statistical testing for claim substantiation. If it is possible to design experiments and collect additional
data, thereby increasing the amount of information, it is possible to have
increased protection against both the type I and type II errors by controlling α and β at lower levels. A large data set provides more information and reduces both α and β.

In claim substantiation, a product is claimed to be better than its
rival with respect to some quantifiable attributes. Often in claims, one
product is claimed to exceed its rival by some specified amount. Thus,
one may be testing a research hypothesis that states that some quantifiable attributes of product A exceed those of product B by a specified
amount, say D. Such a research hypothesis when statistically stated, is
called an *alternative hypothesis H_1*:

H_1: Average attribute of A – average attribute of B > D

The corresponding null hypothesis would be

H_0: Average attribute of A – average attribute of B $\leq$ 0

The prescribed difference D is essentially a "handicap" assigned to
product B. Specification of the difference D in the formulation of the
hypothesis is a valuable technique for substantiating parity or unsurpassed claim.

Statistical Significance and Experimental Significance

Statistical analysis is probabilistic. A statistically significant result may not be of practical significance to the consumers. For example, color in a cosmetic product may have changed over time from its original color. The change may be statistically significant, but not necessarily in the eyes of the consumer. Thus, instead of merely having a statistically significant change, one may need to determine the amount of change that will be perceived as significant by the consumer. This amount of change must be determined by correlating trained panel results with consumer test results.

TYPES OF CLAIMS

According to Smithies (1), claims may be classified by two properties: style and competitive focus. *Style* refers to the statement being made about the advertised brand, the most common being a "distinction" claim, in which a brand claims to be preferred, more efficacious, safer, etc. Another style is a "similarity" claim, which conveys that the advertised product is like the competitor's product in one or more attributes. All products in this category must be tested against the advertised product.

A *competitive focus* claim is a statement being made about the competition against one or more explicitly identified brands or implied brands. For example, the claim may be targeted against an implied brand, i.e., "Preferred over the leading brand," or more broadly against a brand set, i.e., "No leading oil is more absorbent."

In both style and competitive focus, claim statement can be monadic, making no comparison with other products, i.e., a statement of quality, an invitation to try product, or an untargeted claim. An untargeted claim is considered puffery and requires no formal substantiation.

Models for Analysis of Product Performance Data

A useful model for the evaluation of a proposed claim must address the following aspects (2): rationale, objective evidence, subjective evidence, and safety (Fig. 1). The model serves as a guideline that will indicate whether a performance claim is possible and can be substantiated.

COORDINATES OF A COSMETIC CLAIM

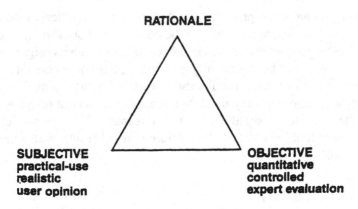

RATIONALE

SUBJECTIVE
practical-use
realistic
user opinion

OBJECTIVE
quantitative
controlled
expert evaluation

Figure 1   Classes of data for claim support (Ref. 2).

A model incorporating these aspects becomes increasingly important in disputed claims.

Rationale

Consumer products contain ingredients that affect the perception that they are desirable. By linking ingredients in a product to experimental results, one can provide a rationale for the claim. Experimental support from allied sciences, such as in vitro studies and model systems, can also provide additional rationale for the claim.

Objective

A claim becomes stronger if its usefulness can be objectively and subjectively determined. An objective measure of product performance is desirable. It can be obtained by studies on test animals, humans, and controlled or model systems. Responses from such studies can be measured by instruments or obtained by trained or expert panels. Indeed, data from trained panels are recognized as objective measures. Descriptive analysis techniques such as Tragon's Quantitative Descriptive Analysis (3) and the Spectrum Method (4), which used a trained panel, can provide objective measures of sensory properties of personal care products.

A descriptive panel undergoes rigid training and calibration as specified by each method. For more information, refer to an ASTM (American Society for Testing and Materials) publication edited by Hootman (5) and to an edited book by Gacula (6) that compiles research work on descriptive sensory analysis, panel training, methods of statistical analyses, and applications to various consumer products and materials.

Subjective

When properly carried out, subjective measures obtained from home-use tests, research guidance panel tests, and the central location consumer tests, among others, may provide useful and acceptable data for claims substantiation.

Safety

Obviously, cosmetics products must be safe and without adverse side effects. The model must address the safety aspects. Safety-related data can be obtained from research guidance panel tests, central location consumer tests, and the various types of laboratory model systems, i.e., in vitro and in vivo tests.

As indicated, a model incorporating these aspects provides a way to deal with conflicts, permits more efficient use of data for the development of truthful claims, and promotes effective communications between parties in disputed situations.

A conceptual model for assessing perception data measuring interdependent attributes is postulated by Wind et al. (7). Briefly, as shown in Fig. 2, this model defines the following:

1. Whether the product performance claim is based on a product attribute or not—i.e., the emphasis is on overall product performance with no focus on product attribute or benefit.
2. Whether the product performance claims focus on a specific attribute or on a set of attributes—i.e., drag, stickiness, residue, spreadability.

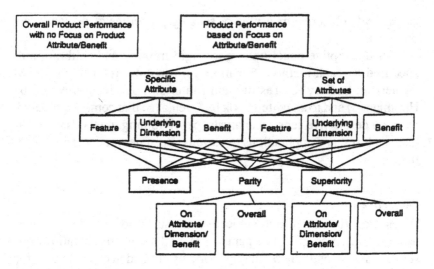

Figure 2   Conceptual model of perceived attribute interdependencies (Ref. 7).

3.   Whether the claim is for a specific attribute or a set of attrib-
     utes and whether it focuses on a feature or a benefit. For
     example, in antibacterial personal care products, the active
     ingredient provides product dimension and the benefit of this
     dimension is cleanliness and safety to the users against bac-
     terial infection.
4.   Whether the claim is merely stating the presence of the
     attribute or also its benefit and whether it suggests a parity or
     superiority against a specific competitor or class of competi-
     tors.
5.   Finally, whether the parity or superiority claim is restricted
     to an attribute and its advertised benefits or to an overall par-
     ity or superiority. For example, in consumer tests, an overall
     preference or overall liking when used as a claim suggests
     that all product attributes contribute, incorporating interde-
     pendence among sensory attributes during product evalua-
     tion by consumers.

   The experimental designs suited for obtaining data for parity and
superiority claims are discussed further on.

Superiority Claims

A superiority claim simply indicates that the product advertised is the best in the market. It is essential that direct product-to-product comparative testing be used for substantiating a superiority claim. An appropriate design for comparing two products at a time is known as the *paired-comparison design*, which is discussed further on. An example of a superiority claim is "Compared to the leading brands, Tropical Isles is unsurpassed as a skin moisturizer and conditioner." As stated before, a claim must be translated into a statistical hypothesis. In order to do this, we must have a well-defined scale on which these products can be scored for comparison. Suppose product A is being compared with a leading brand for claim substantiation. If, on a scale for comparing such products, high scores correspond to superior products, we can formulate two statistical hypotheses such as the following:

H_0: Average score of leading brand $\geq$ average score of product A

H_1: Average score of leading brand $<$ average score of product A

To be able to claim superiority for product A, the null hypothesis must be rejected at, say, the 95% confidence level (5% significance level) in favor of the alternative hypothesis, which states that product A is superior to the leading brand.

Parity Claims

Parity claims are difficult to establish by means of hypothesis testing methodology, because for parity claims the research hypothesis essentially states that the products are equivalent. Using a rating scale, the equivalency is translated as equality of two average scores of the products being compared. However, in the testing of hypothesis methodology, a research hypothesis is not stated as the equality of average scores. Equality of average scores can only be stated as the null hypothesis. A statistical test will either reject the null hypothesis when there is sufficient evidence in support of the alternative or will not reject it. If the null hypothesis is not rejected, it should not be understood that the products are equivalent. Intentionally or otherwise, one can design an experiment to collect insufficient data, lacking in information that leads to a decision not to reject H_0. This decision only means that there is insufficient information to disown the parity claim. It does not mean that parity claim is established with any degree of confidence.

In disputed parity claims, if a proper formulation of hypotheses and a sound design are not used, differences may arise that will be difficult to resolve among parties involved. It is a waste of time to argue about the validity of a claim if the methodology and the design are not carefully employed. As stated above, one can design an experiment with insufficient sample to mask significant difference between products because of the failure of the study to reject the null hypothesis. In his 1935 classic book, *The Design of Experiments*, Fisher (8) wrote the following: "In relation to any experiment, we may speak of this hypothesis as the null hypothesis and it should be noted that the null hypothesis is never proved or established, but it is possibly disproved, in the course of experimentation. Every experiment may be said to exist only in order to give the facts a chance of disproving the null hypothesis."

Therefore, the formulation of hypotheses for a parity claim and their statistical testing must be done in such a way that the decision to reject the null hypothesis amounts to the parity of products. Blackwelder (9) proposed incorporating a specified difference between the scores of products in the formulation of the null and alternative hypotheses. Thus, if the research pertains to establishing a parity claim between product A and the leading brand, one can specify a critical amount D that would be considered a significant difference between product scores if they are really different. In the hypothesis testing terminology, this amounts to formulating the following null and alternative hypotheses:

H_0: Average score of leading brand $\geq$ average score of product A + D

H_1: Average score of leading brand < average score of product A + D

A difference between the leading brand and product A of amount less than D would be understood as saying that product A is at parity with the leading brand. Thus, the statistical decision to reject the null hypothesis would imply a claim for parity.

Example

This example illustrates a computational procedure for testing a null hypothesis with a specified difference to claim product equivalence or parity. In the Table below for illustration for instance, a specified difference was set at $D = 0.25$. In practice, D is close to one standard error of the generally observed sample statistics in consumer testing work. In this example, 10 panelists provided ratings for two products, Tropical Isles and a leading brand. Let L and T denote observations made on the leading brand and Tropical Isles respectively.

Panelist	Leading brand	Tropical Isles	$d = L - T - D$
1	7	5	1.75
2	5	5	–0.25
3	8	6	1.75
4	6	7	–1.25
5	7	6	0.75
6	6	5	0.75
7	7	6	0.75
8	6	6	–0.25
9	7	6	0.75
10	7	5	1.75
Mean	6.60	5.70	0.65
Standard deviation	0.84	0.67	0.99

To claim parity of Tropical Isles with the leading brand, we formulate two statistical hypotheses, as follows: Let μ_d = true mean of L – (true mean of $T + 0.25$). Then

$$H_0: \mu_d \geq 0$$
$$H_1: \mu_d < 0$$

For decision making, a confidence interval can be computed using the formula $\bar{d} \pm k(\text{se})$, where k is a constant corresponding to a chosen confidence level, i.e., 95% and 99% being the most common. In the expression, $\bar{d}$ is the mean of the sample differences d. For a 95% confidence interval, $k = 1.960$, and for a 99% confidence interval, $k = 2.576$, the respective percentiles of the standard normal distribution. For a more accurate interval we would have used the percentiles of the t distribution. The multiplier (se) is the standard error of $\bar{d}$.

In our example, we will use the 99% confidence interval which is:

$$0.65 \pm 2.576(0.31) = 0.65 \pm 0.80$$

i.e., an interval from –0.15 to 1.45. Since the interval includes zero, the two products are judged sensorily at parity with a 99% degree of confidence.

Instead of using the confidence interval to decide between H_0 and H_1, one may use a statistical decision rule such as the following. At a specified significance level α, say 0.01, one may claim parity by rejecting H_0 if $\bar{d} \leq -2.32(se)$. Following this rule, parity would not be claimed if $\bar{d} > -2.32(se)$. We note that (–2.32) is the first percentile of the standard normal distribution, which cuts off 1% area in the left tail leading to a one-sided decision rule: reject H_0 if $\bar{d} < -2.32(se)$. Had it been more appropriate to test $H_0: \mu_d = 0$ against a two-sided alternative $H_1: \mu_d \neq 0$, our rule would be: reject H_0 if $\bar{d} < -2.57(se)$ or $\bar{d} > 2.57(se)$. Thus the use of confidence interval is more consistent with the two-sided tests unless one is willing to use one-sided confidence intervals.

For a decision rule, the probability of its rejecting H_0 is called its *power*. Hence for an optimum rule, its power must tend to 1.0 as the true mean difference μ_d tends to be further away from H_0 and to 0 otherwise. The power of a decision rule is $(1 - \beta)$, where β is the probability of type II error. In our example, β is given by

$$\beta = \Phi[2.32 + (\mu_d \sqrt{n}/0.99)]$$

In the above expression, $\Phi(x)$ refers to the cumulative probability of the standard normal curve. Note that the expression for x in $\Phi(x)$ depends upon μ_d and n, the sample size. Hence not only β but also the power = $1 - \beta$ depends upon μ_d and n. As one would expect from an optimum rule, its power must increase with n and μ_d as it moves away from H_0 in the domain of the alternative hypothesis. This is illustrated in Figure 3. The three power curves have been plotted for $n = 10, 20$, and 30. Note that all the three curves increase rapidly from 0.01 when $\mu_d = 0$ toward 1.0 as the alternative hypothesis becomes more and more pronounced, with μ_d negative and large in magnitude. However, the three curves rise to 1.0 at a different rates; the curve with large n rises at a faster rate. This is because the power of a rule increases with sample size n; the larger

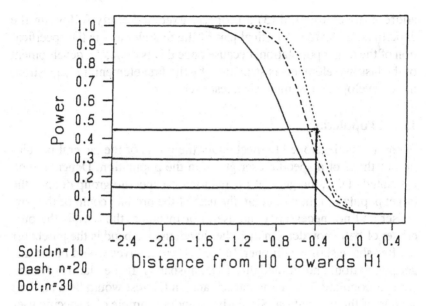

Solid;n=10
Dash; n=20
Dot;n=30

Figure 3 Power curves for $n = 10, 20, 30$.

the n, the more power. Note the power values from the three curves of the same rule except different values of n. For instance, if the alternative is true with a true difference $\mu_d = -0.4$, the decision rule would reject the null hypothesis H_0 with probabilities 0.15, 0.30, and 0.46 corresponding to $n = 10, 20$, and 30, respectively.

We conclude this section with one important observation. If in a claims substantiation investigation it is desired to control the probability of type I error at a specified level and to have the desired power, one would need to know the sample size n. How large a sample must be collected to have a rule with specified power and type I probability? There are many sources one can consult for determining sample size n, such as Kraemer and Thiemann (10).

EXPERIMENTAL DESIGNS FOR CLAIM SUPPORT

There are three important elements in the development of a strong product claim: (1) a clearly stated claim, (2) a good experimental design to address the claim, and (3) a properly executed study following the experimental design. A critical part of the first element is the specification of the target population, because once this is done, the development of the last two elements is structured by the first element, i.e., questionnaire development, sample size, test execution.

Target Population

A product is developed to meet either the needs of the general population or those of a specific user group in the population. Depending on the stated claim, the general population or a specific group defines the target population. In particular, the user of the product could be the purchaser and not necessarily the user. For instance, the wife is the purchaser of baby powder. On the other hand, the husband is the purchaser of after-shave skin conditioner. In the first case, wives would be the target population, and in the latter case, husbands. If the claim is for the general population, then the participants in the test would be a random sample of the population. Similarly, a random sample of a specific user group should be used in the study. Sampling techniques are not covered in this chapter, but these techniques are widely published (i.e., Refs. 11–13).

Questionnaire Design

In gathering consumer data for claim substantiation, it is important that the product attributes related to the claim be included in the questionnaire. For example, if "soft" and "smooth" are sensory attributes claimed for the product, then these attributes must be included in the questionnaire in the form of intensity and/or hedonic (like/dislike) questions.

How many attribute questions the questionnaire should include is often a difficult decision to make in questionnaire development. If a product has undergone a series of descriptive sensory analyses, this should provide the appropriate number of attributes for inclusion. Briefly, descriptive analysis is a sensory methodology that provides quantitative descriptions of products based on the perceptions of a group of qualified subjects. It is a total sensory description, taking into account all sensations perceived—visual, auditory, olfactory, kinesthetic, and so on—when the product is evaluated (3). In practice, the desirable number of attributes has ranged from 10 to 15.

Another aspect of questionnaire development is the choice of rating scales. The best-known scale used in consumer testing is the 9-point hedonic scale (1 = dislike extremely, 5 = neither like nor dislike, 9 = like extremely) developed in 1947 at the Quartermaster Food and Container Institute for the U.S. Armed Forces (14). This is the most extensively studied of rating scales and, as a result, is the most reliable one for acceptance/preference measurement. Information on questionnaire development is widely available (3,4,15). (See Ref. 16 for the calculation of sample size in consumer tests.)

Paired Comparison

The paired comparison is the most powerful design to support almost all types of product claims. For instance, a 1994 NAD statement says that "Direct product-to-product comparative testing is a significant, if not essential element of the substantiation of a superiority claim" (17).

The statistical analysis of paired-comparison design is simple and meets all the essential statistical assumptions; the test is simple to execute for both the experimenter and the panelists, and the evaluation of two products by a single panelist fits nicely into the classic paired-comparison situation (i.e., right/left sides of biological materials).

The general idea of the paired-comparison design is to form homogeneous pairs of like units so that comparisons between units of a pair measure differences due to treatments rather than units. This arrangement leads to dependency between observations on units of the same pair. This situation can be extended to sensory and consumer testing. Panelists compare two products by assigning a score. Thus the scores X_{1i} and X_{2i}, $i = 1, 2, \ldots, n$ panelists, being made by the ith panelist, are not independent and constitute a pair (X_{1i}, X_{2i}). Observations X_{1i} and X_{2i} are correlated because panelists who are high raters tend to give high scores to both experimental units (product samples) being compared. Likewise, low raters tend to give low scores to both experimental units. Since the score assigned to the pair of units came from the same panelists, the differences $d_i = X_{1i} - X_{2i}$ measure only the differences between units independently of how panelists used the rating scale. The statistical assumption in the analysis is that the difference d_i is independent and normally distributed; in most cases this assumption is satisfied in practice. Furthermore, the common problem of correlation of ratings among panelists becomes irrelevant, since one is now dealing with differences d_i.

Example

For the purpose of illustration, consider the consumer data in Table 3 using $n = 10$ panelists. The purpose of this illustration is to show a simple SAS (18) program for analyzing the data. The data were tabulated following the sequence that panelists evaluated the products according to the questionnaire. For example, panelist 1 evaluated the product in the order (A,B) for attributes Q1 to Q9. Note that Q9 is an overall preference question, where a 1 is assigned to the most preferred sample and a 2 to the least preferred. The remaining attributes were evaluated on a 9-point hedonic scale.

Table 4 shows the SAS code (program) used to analyze the data in Table 3. The SAS code was purposely written to illustrate the most useful statistics in consumer data: calculations of descriptive statistics (PROC MEANS), construction of chart (PROC CHART), and frequency distribution (PROC FREQ). Charts and frequencies for the other

Table 3 Sensory Data for Two Products A and B on Nine Attributes

Pan[a]	Order	Q1	Q2	Q3	Q4	Q5	Q6	Q7	Q8	Q9
1	A,B	7,6	6,6	6,6	7,6	5,5	6,5	8,7	5,6	1,2
2	B,A	6,6	5,6	5,6	6,6	5,7	6,5	7,6	7,7	2,1
3	A,B	7,6	5,6	6,5	5,5	6,5	7,5	5,5	7,6	1,2
4	B,A	5,6	5,6	5,7	6,6	5,7	6,5	5,6	5,6	2,1
5	A,B	5,5	5,6	6,5	5,5	5,5	5,5	6,5	7,5	1,2
6	B,A	5,7	5,7	6,6	5,5	6,6	6,6	5,7	6,7	2,1
7	A,B	5,6	7,6	6,6	7,6	6,6	6,6	6,6	6,6	1,2
8	B,A	6,8	5,6	6,7	5,7	6,7	6,7	5,7	6,7	1,2
9	A,B	7,5	6,6	5,5	6,6	6,7	5,6	7,5	7,6	1,2
10	B,A	6,7	7,7	7,6	7,7	6,7	7,6	6,7	6,7	2,1

[a]Pan = panelist.

Table 4   SAS Code for Paired Comparison Analysis

```
*PROGRAM PAIRED;
%LET TITLE=TABLE 3 SAMPLE DATA;
TITLE"&TITLE";
DATA LOTION;
FILENAME IN 'A:SAMPLE1';
INFILE IN;

INPUT PAN OR1 $ OR2 $ @@;

LABEL X1A='OVERALL LIKING'
      X2A='FRAGRANCE APPROPRIATENESS'
      X3A='FRAGRANCE PLEASANTNESS'
      X4A='LATHER LIKING'
      X5A='CLEAN RINSING'
      X6A='SKIN SOFTNESS AFTER DRYING'
      X7A='SKIN SMOOTHNESS AFTER DRYING'
      X8A='MOISTURIZES SKIN'
      X9A='OVERALL PREFERENCE'

      X1B='OVERALL LIKING'
      X2B='FRAGRANCE APPROPRIATENESS'
      X3B='FRAGRANCE PLEASANTNESS'
      X4B='LATHER LIKING'
      X5B='CLEAN RINSING'
      X6B='SKIN SOFTNESS AFTER DRYING'
      X7B='SKIN SMOOTHNESS AFTER DRYING'
      X8B='MOISTURIZES SKIN'
      X9B='OVERALL PREFERENCE';

IF OR1='A' THEN DO; OR2='B';
INPUT X1A X1B X2A X2B X3A X3B X4A X4B X5A X5B X6A X6B X7A
   X7B X8A X8B X9A X9B;
END;

IF OR1='B' THEN DO; OR2='A';
INPUT X1B X1A X2B X2A X3B X3A X4B X4A X5B X5A X6B X6A X7B
   X7A X8B X8A X9B X9A;
END;

DATA LOTION;
SET LOTION;

   DIFF1=X1A–X1B;
```

Continues

Table 4 Continued

```
    DIFF2=X2A–X2B;
    DIFF3=X3A–X3B;
    DIFF4=X4A–X4B;
    DIFF5=X5A–X5B;
    DIFF6=X6A–X6B;
    DIFF7=X7A–X7B;
    DIFF8=X8A–X8B;
    DIFF9=X9A–X9B;
    RUN;

PROC PRINT DATA=LOTION;
    TITLE1"&TITLE";
    TITLE2"RAW DATA";
    RUN;

DATA LOTION;
SET LOTION;

PROC MEANS MEAN N STD STDERR MAXDEC=2;
    VAR X1A X2A X3A X4A X5A X6A X7A X8A X9A;
    TITLE1"&TITLE";
    TITLE2"DESCRIPTIVE STATISTICS FOR PRODUCT A";
    RUN;

PROC MEANS MEAN N STD STDERR MAXDEC=2;
    VAR X1B X2B X3B X4B X5B X6B X7B X8B X9B;
    TITLE1"&TITLE";
    TITLE2"DESCRIPTIVE STATISTICS FOR PRODUCT B";
    RUN;

PROC MEANS MEAN N STD T PRT MAXDEC=4;
    VAR DIFF1–DIFF9;
    TITLE1"&TITLE";
    TITLE2"MEAN DIFFERENCES (A–B) AND SIGNIFICANCE TEST";
    RUN;

PROC CHART;
    HBAR X1A / DISCRETE;
    TITLE1"&TITLE";
    TITLE2"OVERALL LIKING CHART FOR PRODUCT A";
    RUN;
```

Table 4 Continued

```
PROC CHART;
  HBAR X1B / DISCRETE;
  TITLE1"&TITLE";
  TITLE2"OVERALL LIKING CHART FOR PRODUCT B";
  RUN;

PROC FREQ;
  TABLE X9A*X9B;
  TITLE1"&TITLE";
  TITLE2"PRODUCT A=X9A PRODUCT B=X9B 1=PREFERRED
    2=NOT PREFERRED";
  RUN;
```

attributes can easily be done using the core program in Table 4. If one runs the SAS program, one finds significant differences at the indicated probability level (p level) in favor of product A for the following attributes:

DIFF1 (overall liking)	p level = 0.0187
DIFF7 (skin softness after drying)	p level = 0.0187
DIFF8 (moisturizes skin)	p level = 0.0248
DIFF9 (overall preference)	p level = 0.0031

In the overall preference, there were nine panelists who indicated preference for product A against one panelist for product B. A simple χ^2 test can also be used to analyze the overall preference data.

Randomized Complete Block Design

For reasons of cost, time, and other business constraints, one must conduct a consumer test with more than two products for evaluation by panelists at the same time. In this situation, the randomized complete block design (RCBD) is used for claim substantiation. The statistical model for describing an observation is

$$Y_{ij} = \mu + A_i + B_j + e_{ij}$$

where Y_{ij} = the observed rating for the ith product given by the jth panelist; μ = the grand mean; A_i = the effect of the ith product; B_j = the effect of the *j*th panelist; and e_{ij} = random errors assumed to be independently and normally distributed, with mean zero and variance σ_e^2. In this model, the effect of panelist-to-panelist variation is removed from the random errors e_{ij}, making the test of significance more sensitive.

Table 5 contains the SAS program for analyzing an RCBD. As input, the program reads the columns in the following order: panelist, product, attributes (Q1 to Q8). For example:

Pan	Prod	Q1	Q2		Q8
1	A	7	6	...	5
1	B	6	6	...	6
1	C	5	4	...	7
etc.					

In this statistical model, there were two main effects—product and panelist—which is equivalent to a single-factor repeated-measures design. When more than two factors are involved, the repeated-measures design is recommended. A common main effect (factor) in sensory and consumer tests is time. When panelists evaluate products over time, correlation (dependency) among the ratings is present and may violate the assumptions in the statistical analysis. To account for this dependency, the repeated-measures design or other multivariate methods of analysis are used. These subjects are not discussed here; interested readers are referred to Morrison (19), Cody and Smith (20), Girden (21), and Littell et al. (22).

However, in most consumer testing claim studies, the statistical analysis from the RCBD or the single-factor repeated-measures design is sufficient. Also, the SAS code in Table 5 can easily be expanded to include demographics, product usage information, and so on.

CONCLUDING REMARKS

In this chapter, we have tied the importance of statistical experimental design to consumer tests for supporting claim substantiation. In particular, the formulation of statistical research hypotheses is discussed and its importance in parity claims reviewed. The use of a paired-comparison design is recommended for claims substantiation. The importance

Table 5 SAS Code for Randomized Complete Block Analysis

```
*PROGRAM RCBD;
%LET TITLE=TABLE 5 SAMPLE DATA;
TITLE"&TITLE";
FILENAME IN "A:SAMPLE";

DATA LOTION;
INFILE IN;
INPUT PAN PROD $ Q1-Q8;
LABEL Q1='OVERALL LIKING'
      Q2='FRAGRANCE APPROPRIATENESS'
      Q3='FRAGRANCE PLEASANTNESS'
      Q4='LATHER LIKING'
      Q5='CLEAN RINSING'
      Q6='SKIN SOFTNESS AFTER DRYING'
      Q7='SKIN SMOOTHNESS AFTER DRYING'
      Q8='MOISTURIZES SKIN';

DATA LOTION;
SET LOTION;

PROC SORT;
   BY PROD;
   RUN;

PROC PRINT;
   TITLE1"&TITLE";
   TITLE2"RAW DATA";

%MACRO A;
   %DO I=1 %TO 8;
   DATA;
   SET LOTION;

PROC MEANS N MEAN STD STDERR MAXDEC=2;
   VAR Q&I;
   BY PROD;
   TITLE1"&TITLE";
   TITLE2"DESCRIPTIVE STATISTICS FOR TWO LOTION PROD-
   UCTS";

PROC GLM;
   CLASS PROD;
```

Continues

Table 5   Continued

```
    MODEL Q&I = PAN PROD;
    MEANS PROD/BON;

%END;
%MEND A;
%A;
RUN;

DATA LOTION;
SET LOTION;

    PROC FREQ;
    TABLES PROD*(Q1 Q2 Q3);
    TITLE1"&TITLE";
    TITLE2"FREQUENCY DISTRIBUTION";
    RUN;

PROC CHART;
    HBAR Q1 / DISCRETE;
    BY PROD;
    TITLE1"&TITLE";
    TITLE2"OVERALL LIKING CHART";
    RUN;
```

---

of understanding the power of a statistical test and its relationship to sample size to provide a claim that can withstand rigorous scrutiny was emphasized.

REFERENCES

1. Smithies RH. Resolving advertising disputes between competitors. Food Technol 1994; August: 68.
2. Smithies RH. Model for analysis of product performance data: Application to cosmetics and drugs. NAD Workshop IV Proceedings, Product Performance Tests, Rye, New York. New York: Council of Better Business Bureaus, 1992.
3. Stone H, Sidel JL. Sensory Evaluation Techniques. San Diego, CA: Academic Press, 1993.

4. Meilgaard M, Civille GV, Carr BT. Sensory Evaluation Techniques. Boca Raton, FL: CRC Press, 1991.
5. Hootman RC, Ed. Descriptive Analysis Testing. ASTM Manual Series: MNL 13. Philadelphia, PA: ASTM, 1992.
6. Gacula MC Jr, ed. Descriptive Sensory Analysis in Practice. Trumbull, CN: Food & Nutrition Press, 1997.
7. Wind Y, Schmittlein DC, Shapiro S. Attribute interdependencies in product performance claims: Truth and consequences. NAD Workshop IV Proceedings, Product Performance Tests, Rye, New York. New York: Council of Better Business Bureaus, 1992.
8. Fisher RA. The Design of Experiments. New York: Hafner, 1960.
9. Blackwelder WC, Chang MA. Sample size graphs for proving the null hypothesis. Contr Clin Trials 1984; 5:97–105.
10. Kraemer HC, Thiemann S. How Many Subjects?: Statistical Power Analysis in Research. Thousand Oaks, CA: Sage, 1987.
11. Snedecor GW, Cochran WG. Statistical Methods. Ames, IA: Iowa State University Press, 1967.
12. Cochran WG. Sampling Techniques. New York: Wiley, 1963.
13. Kish L. Survey Sampling. New York: Wiley, 1965.
14. Peryam DR, Girardot HF. 1952. Advanced taste-test method. Food Eng 1952; 24:58–61, 194.
15. Moskowitz H. Cosmetic Product Testing: A Modern Psychophysical Approach. New York: Marcel Dekker, 1984.
16. Gacula M C Jr, Singh J. Statistical Methods in Food and Consumer Research. San Diego, CA: Academic Press, 1984.
17. NAD Case Reports. New York: Council of Better Business Bureaus, July 1994:81.
18. SAS/STAT User's Guide. Cary, NC: SAS Institute Inc, 1990.
19. Morrison DF. Multivariate Statistical Methods. New York: McGraw-Hill, 1967.
20. Cody RP, Smith JK. Applied Statistics and the SAS Programming Language. New York: North-Holland, 1987.
21. Girden ER. ANOVA: Repeated Measures. Newbury Park, CA: Sage, 1992.
22. Littell RC, Milliken GA, Stroup WW, Wolfinger RD. SAS System for Mixed Models. Cary, NC: SAS Institute Inc, 1996.

Index

Printed in the United States
by Baker & Taylor Publisher Services